AF761221

Good Earths

Frontiers of Asia Series

The Social Ecology of Tropical Forests
Wil De Jong, Lye Tuck-Po and Abe Ken-ichi

Good Earths: Regional and Historical Insights into China's Environment
Abe Ken-ichi and James E. Nickum

Good Earths

Regional and Historical Insights into China's Environment

Edited by

Abe Ken-ichi
James E. Nickum

Kyoto University Press

TRANS PACIFIC PRESS

First published in 2009 jointly by:

Kyoto University Press
Kyodai Kaikan
15-9 Yoshida Kawara-cho
Sakyo-ku, Kyoto 606-8305, Japan
Telephone: +81-75-761-6182
Fax: +81-75-761-6190
Email: sales@kyoto-up.gr.jp
Web: http://www.kyoto-up.gr.jp

Trans Pacific Press
PO Box 120, Rosanna, Melbourne
Victoria 3084, Australia
Telephone: +61 3 9459 3021
Fax: +61 3 9457 5923
Email: info@transpacificpress.com
Web: http://www.transpacificpress.com

Copyright © Kyoto University Press and Trans Pacific Press 2009

Distributors

Asia and the Pacific
Kinokuniya Company Ltd.

Head office:
38-1 Sakuragaoka 5-chome, Setagaya-ku, Tokyo
156-8691, Japan
Telephone: +81 (0)3 3439 0161
Fax: +81 (0)3 3439 0839
Email: bkimp@kinokuniya.co.jp
Web: www.kinokuniya.co.jp

Asia-Pacific office:
Kinokuniya Book Stores of Singapore Pte., Ltd.
391B Orchard Road #13-06/07/08
Ngee Ann City Tower B
Singapore 238874
Telephone: +65 6276 5558
Fax: +65 6276 5570
Email: SSO@kinokuniya.co.jp

Japan
Kyoto University Press
Kyodai Kaikan
15-9 Yoshida Kawara-cho
Sakyo-ku, Kyoto 606-8305
Telephone: (075) 761-6182
Fax: (075) 761-6190
Email: sales@kyoto-up.gr.jp
Web: http://www.kyoto-up.gr.jp

UK and Europe
Asian Studies Book Services
Franseweg 55B, 3921 DE Elst, Utrecht,
The Netherlands
Telephone: +31 318 470 030
Fax: +31 318 470 073
Email: info@asianstudiesbooks.com
Web: http://www.asianstudiesbooks.com

USA and Canada
International Specialized Book
Services (ISBS)
920 NE 58th Avenue, Suite 300
Portland, Oregon 97213-3786
USA
Telephone: (800) 944-6190
Fax: (503) 280-8832
Email: orders@isbs.com
Web: http://www.isbs.com

Australia and New Zealand
UNIREPS
University of New South Wales
Sydney, NSW 2052
Australia
Telephone: +61(0)2-9664-0999
Fax: +61(0)2-9664-5420
Email: info.press@unsw.edu.au
Web: http://www.unireps.com.au

All rights reserved. No production of any part of this book may take place without the written permission of Kyoto University Press or Trans Pacific Press.

ISBN 978-1-920901-35-6

Contents

Tables

Figures

Photographs

Acknowledgments

This volume arises from the papers prepared for the International symposium entitled 'Impacts on *The Good Earth*: Recent Environmental History in China', held in Kyoto, January 23–25, 2004. The symposium was organized and co-sponsored by the Japan Center for Area Studies, National Museum of Ethnology, Research Institute for Humanity and Nature (RIHN), and the Center for Southeast Asian Studies (CSEAS), Kyoto University. The chapter by Mehtonen was added later. We would like to especially thank Noriko Iizuka for her indispensable coordinating of all the various activities necessary to bring the manuscripts from symposium drafts to finalized published chapters, and Miriam Riley, for her painstaking and careful editing.

Finally we are most grateful to the Japan Society for the Promotion of Science (JSPS) for a Grant-in Aid for Publication of Scientific Research Results, which enabled the publication of this volume.

Contributors

ABE Ken-ichi

Abe Ken-ichi is professor at the Research Institute for Humanity and Nature of the National Institute for the Humanities, Japan. His areas of specialization are environmental anthropology and integrated area studies. He has written and edited many publications, including *Extreme Conflict and Tropical Forests* (Springer, 2007); *The Social Ecology of Tropical Forests: Migration, Population and Frontier* (Kyoto University Press and Trans Pacific Press, 2006); and *The Political Ecology of Tropical Forests in Southeast Asia: Historical Perspectives* (Kyoto University Press and Trans Pacific Press, 2003).

Brian BELCHER

Brian Belcher is director of the Centre for Non-Timber Resources (CNTR) of the Royal Roads University. The overall focus of his work is on understanding and improving the role and potential of natural resources to meet development and environmental objectives. His numerous publications include *China's Forests: Global Lessons from Market Reforms* (ed.) (Resources for the Future Press, 2006); 'NTFP commercialization: A reality check', *Development Policy Review*, 25(3) (2007); and 'Global patterns and trends in the use and management of commercial NTFPs: Implications for livelihoods and conservation', *World Development*, 33(9) (2005).

CANG Ming

Cang Ming is professor in the Department of History of the Central University for Nationalities in Beijing. His areas of specialization are the history and cultures of the minorities of South China. His numerous publications include (all in Chinese): *The History of Immigration in Yunnan's Border Areas* (National Press, 2005); *Study of the Culture of Yunnan Ethnic Migration* (Yunnan National Press, 1997); and *Huang Yao Ancient Town* (National Press, 2007).

CHEN Yong

Chen Yong is professor at Sichuan University and director of the Institute of Environment and Regional Development. His areas of specialization are human ecological approaches to rural settlement, interdisciplinary studies of population, environment and sustainability, and environmental economic evaluation. He has co-authored six books and published over forty journal papers.

FU Maoyi

Fu Maoyi is an award-winning bamboo researcher at the Research Institute of Subtropical Forestry, of the Chinese Academy of Forestry, where he was formerly director. His many works in both English and Chinese include *A Compendium of Chinese Bamboo* (China Forestry Publishing House, 1994); 'The relationship between forest research and forest management in China: Analysis of four leading Chinese forestry journals', *International Forestry Review* (2004); and 'Bamboo forestry in China: Toward environmentally friendly expansion', *Journal of Forestry* (2001).

HAN Wei

Han Wei was a student at the College of Environmental Sciences, Peking University while working for this project.

William F. HYDE

William Hyde, retired from Duke and Virginia Tech Universities, maintains associations with Gothenburg University in Sweden and Beijing Forestry University. He has written widely on natural resource and environmental economics, especially with regard to rural natural environments. His books include *China's Forests: Global Lessons from Market Reforms* (with Brian Belcher and Jintao Xu, Resources for the Future, 2003); *The Economics of Forestry and Rural Development* (with Gregory Amacher, Michigan 2000); and *Timber Supply and Land Allocation* (Johns Hopkins, 1957).

Leiwen JIANG

Leiwen Jiang is senior demographer at Population Action International, adjunct professor at Brown University and Peking University, and guest researcher at the International Institute for Applied System Analysis and the China Academy of Sciences. His research focuses on population and environmental interactions, particularly population and climate change modeling. He is the author of *Population and Sustainable Development in China* (Thela Thesis, 1999), as well as many journal articles.

KOBAYASHI Tatsuaki

Kobayashi Tatsuaki is professor at the College of Horticulture, Chiba University. His specialty is Landscape and Restoration Ecology. His many publications include *Wild Azalea Whispers—Socio-ecology of SATOYAMA in Boso Peninsula* (Chiba-nippo-sha, 2007); *Handbook of Revegetation for Biodiversity Conservation* (ed.) (Chijin-shokan, 2006); and 'Socio-ecological analysis of desertification in the Mu-Us Sandy Land with satellite remote sensing' *Landscape and Ecological Engineering*, 2005.

LI Tianhong
Li Tianhong is an associate professor in the College of Environmental Sciences, Peking University. He has published many journal articles on water resource problems, especially using techniques of remote sensing and GIS.

LIAO Jianhua
Liao Jianhua is a PhD candidate at the Institute of Geographic Studies, China Academy of Sciences.

LIU Dachang
Liu Dachang is senior conservation scientist at The Nature Conservancy China Program. His academic areas include forestry, natural resources management, and development studies (focusing on rural development). His many publications include 'Policies and programme to make decentralization effective: A case study from China', in *Lessons from Forest Decentralization* (Earthscan, 2008); *Rehabilitation of Degraded Forests to Improve Livelihoods of Poor Farmers in South China* (ed.) (Center for International Forestry Research, 2003); and 'Tenure and management of non-state forests in China since 1950: A historical review', *Environmental History*, 6(2) (2001).

LIU Hua
Liu Hua is associate researcher at The Research Institute of Forest Resources Information Techniques, CAF, China. Her area of specialization is application research of 3S technique in forestry. Her publications include *Research on Dynamic Change of Bamboo Forest Using Remote Sensing Data in Fuyang* (Forest Research, 2008); and *Monitoring Techniques of Sand and Dust Storms* (Forestry Publishing Company, 2008).

Robert B. MARKS
Robert B. Marks is Richard and Billie Deihl Professor of History at Whittier College where he has taught East Asian and Chinese history since 1978. He is the author of several books, most recently *Tigers, Rice, Silk and Silt: Environment and Economy in Late Imperial South China* (Cambridge 1998), and *The Origins of the Modern World: A Global and Ecological Narrative*

from the Fifteenth to the Twenty-first Century (Rowman and Littlefield 2006). He won the Aldo Leopold Award for best article for 'Commercialization without capitalism: Processes of environmental change in South China, 1550–1850', *Environmental History*, (1996).

Katri MEHTONEN

Katri Mehtonen (formerly Makkonen) is Environmental Consultant at Ramboll Whitbybird, and PhD candidate at Water Resources Laboratory of Helsinki University of Technology (TKK). Her areas of specialization are integrated water resources and river basin management, management of international and transboundary rivers and the interconnections between water, politics and economics. She published 'China's role in integrated water resources management in South and Southeast Asia', in O. Varis, A.K. Biswas and C. Tortajada, (eds) *Integrated Water Resources Management in South and South-East Asia* (Oxford University Press, 2005).

Benoit MERTENS

Benoit Mertens was a CIFOR researcher at the time of writing the chapter. He has published extensively on spatial analysis of deforestation processes, including 'Spatial patterns and processes of bamboo expansion in Southern China', *Applied Geography*, (2008).

NAKAWO Masayoshi

Nakawo Masayoshi is Executive Director of the National Institutes for the Humanities, Japan. He does interdisciplinary research on environmental issues in arid and semi-arid regions, especially central Eurasia, by examining the history of interactions between humans and nature. His many publications include *Snow and Ice Science in Hydrology* (Nagoya University and UNESCO), and *Studies in Humanity and Environment of Khara Khoto*, Proceedings of the International Symposium on the Humanity and Environment of Khara Khoto Region (Renmin University of China Press). He has also produced a series of DVDs including *Where did the oasis water go?*

James E. NICKUM

James E. Nickum is professor at Tokyo Jogakkan College and Japan Director of the Asian Water and Resources Institute. His area of specialization is institutional economics, focusing on water. His many publications include *Environmental Management, Poverty Reduction and Sustainable Regional Development* (ed.) (Greenwood Press, 2001); 'Same longitude, different latitudes: Institutional change in urban water in China, North and South',

Environmental Politics, April (2006); and 'Government office and dense clouds: The large dams that command Beijing's heights' (forthcoming).

Lesley POTTER

Lesley Potter is Associate Professor, Visiting Fellow in Human Geography at the Australian National University. Her main research interests include commodities, governance, and decentralisation, especially in Indonesia. Her recent publications include: 'Production of people and nature, rice and coffee: The Semendo people in South Sumatra and Lampung', in N. Peluso and J. Nevins (eds), *Taking Southeast Asia to Market: Commodities, People and Nature in a Neoliberal Age* (2008); 'Governance, tenure and equity in Asia-Pacific forests' in C. Colfer and D. Capistrano (eds), *Forest Governance and Decentralisation in Asia and the Pacific* (2008); and 'Resource periphery, corridor, heartland: contesting land-use in the Kalimantan/Malaysia borderlands', *Asia-Pacific Viewpoint* (forthcoming, April 2009).

Manuel RUIZ-PEREZ

Manuel Ruiz-Perez is associate professor at the department of Ecology, Autonomous University of Madrid, Spain. His area of specialization is natural resources management, with a special interest in forest conservation and management, and forest-based rural livelihoods. Among his prolific publications are *Current Issues in Non-Timber Forest Products Research* (1996); 'Looking through the bamboo curtain: An analysis of the changing role of forest and farm income in rural livelihoods in China', *International Forestry Review* (2004); and 'External influences on and conditions for community logging management in Cameroon', *World Development*, (in press, 2008).

TONG Yufen

Tong Yufen is professor at the Institute of Population Research at Xinjiang University. She received her PhD in population economics from Peking University, with expertise on the systems dynamics analysis of population–environment interactions. She is the author of *Population and Sustainable Development—Theoretical Methods and Practice*, and many other books and journal articles.

TSUI Yanhu

Tsui Yenhu (CUI Yanhu) is professor at the Institute of Social-cultural Anthropology, Xinjiang Normal University, China. His area of specialization is eco-anthropology and area development studies. His many publications include 'A comparative study of the attitudes of the peoples of pastoral areas of inner Asia towards their environment', and 'The development of social organizations in the pastoral areas of North Xinjiang and their changing relationship with the environment', both in *Culture and Environment in Inner Asia: Society and Culture* (White Horse Press, 1997); and *Population, Resources, Subsistence Systems and Eco-environmental Changes in the Grassland of North Xinjiang,* (Xinjiang People's Publisher, 2002).

WANG Yiqian
Wang Yiqian is professor and deputy director of the Institute of West China Development, Sichuan University. His areas of specialization are poverty alleviation, environment, migration and development with a focus on western China. He has produced numerous area-based research reports for local governments and published over seventy research papers.

XIE Jinzhong
Xie Jinzhong is Associate Professor, The Research Institute of Subtropical Forestry, Chinese Academy of Forestry

Jintao XU
Jintao Xu is a professor of natural resource economics at Peking University. His research ranges from forest tenure reform and industrial pollution control policy to carbon sequestration from agricultural land use changes. He has written papers on assessments of China's Sloping Land Conversion Program, pulp and paper mills' productive and environmental performance and state forest reforms, published in journals such as *Land Economics*, *Ecological Economics* and the *International Forestry Review*. Currently he serves as an associate editor of *Environment and Development Economics.*

YAN Tan
Yan Tan is a Research Fellow at the National Institute of Labour Studies at Flinders University, Australia. Her areas of specialty are migration and resettlement, environmental change and adaptation. Her many publications include *Resettlement in the Three Gorges Project* (Hong Kong University Press, 2008); 'Environmental migration and sustainable development in the upper reaches of the Yangtze River', *Population and Environment*, July (2004); and 'Linking migration responses to environmental degradation and change in West China', IOM Research Series (forthcoming).

YANG Xiaosheng

Yang Xiaosheng is Associate Professor, The Research Institute of Subtropical Forestry (RISF), Chinese Academy of Forestry (CAF). His specialization is the economics of forest resources, especially economic evaluation, ecological assessment and policy instrument choice linked to social economy. His publications include 'Evaluation indicators of modern forestry', in *China Modern Forestry* (China Forestry Publishing House, 2008) and 'Evaluation services: A case study in the economics and ecosystems of *moso* bamboo stands, Huzhou City, Zhejiang Province', *Journal of Bamboo Research* (2007).

YIN Shaoting

Yin Shaoting is professor at The Ethnic Group Research Institute of Yunnan University, China. His areas of specialization are ecological anthropology, eco-environmental history and cultural heritage protection. He has written and edited many publications, including *The Ethnic, Cultural and Ecological Villages: An Exploration in China's Applied Anthropology* (Yunnan University Press, 2008); *Cultural Ecology and Material Culture* (Yunnan University Press, 2007); and *Mountain Fire Fading Away: Swidden Agriculture in Anthropological Perspective* (The Peoples' Publishing House of Yunnan, 2008).

ZHAO Zhijie

Zhao Zhijie is an associate professor in the College of Environmental Sciences, Peking University. He specializes in environment integrated assessment and land analysis. His research received a Science and Technology Advancement award from the China Ministry of Education.

Preface: Earth, Wood, Water and Grass: China's Environmental Margins

Abe Ken-ichi and James E. Nickum

Everyone knows that China is large in both space and time. Yet it is easy to slip into the habit of thinking of China as having 'an' environment, encompassing all within its international boundaries; or of seeing the impact of recent economic growth and urbanization as somehow uniquely and uniformly degrading to that environment, producing a 'Bad Earth' (cf. Smil 1984). By so generalizing, we lose sight of regional diversity and the complexity of the interactions between human and natural systems, both historically and in the present day.[1]

In fact, China has a wide range of natural environments: from deserts to tropical forests, from the world's highest mountains to some of its greatest river deltas. Similarly, its human communities range from nomads to farmers to urban dwellers, from the long settled to the migratory, from the Han majority with all its sub-ethnicities to a broad range of present and historical minority ethnicities who still populate half the territory of the People's Republic of China. The geographically expansive ethnic margins also, not coincidentally, happen to be the locus of many of China's most stressed ecological peripheries.

Environmental issues are often perceived as contemporary problems, and some are, such as the consequences of the chemicalization of agriculture. Many concerns, however, such as land degradation and desertification, have deep historical roots. It is not possible to understand the present, or to find solutions to environmental problems, without a solid understanding of historical processes. Here, the vast length of China's history and the wealth of surviving records are unique resources whose importance extends far beyond that country's boundaries.

The contributions that form this volume arise out of The Eighth International Symposium of the Joint Research Project on State, Nation, and Ethnic Relations, titled 'Impacts on The Good Earth: Recent Environmental History in China', convened by the Japan Center for Area Studies in cooperation with the Research Institute for Humanity and Nature (RIHN) and the Center for Southeast Asian Studies, Kyoto University (CSEAS), and held in Kyoto on 23–25 January, 2004. This symposium sought to bring together largely empirical studies aimed at understanding regional changes within China over time with a view towards improving our ability to assess specific, regional environmental issues. One-third of the chapters are by Chinese scholars, many of them from leading regional universities. Because environmental margins in China are often also ethnic 'margins', there is some but not exclusive emphasis on minority nationalities. Pearl Buck (2004[1931]) long ago made the 'Good Earth' a synonym for China, and specifically for a romanticized vision of people then living in the environmentally problematic North China plain region. We borrowed Buck's term for the symposium and the title of this book.

Our focus is on those things that cover and nourish these 'good earths' in a healthy environment: the land itself, trees, water and grasses (including bamboo). Those familiar with Chinese cosmology will note the allusion to the five elements (earth, wood, water, fire and metal). Fire and metal we leave to others, while noting that the ancients may have underrated grasses in their cosmology, although they certainly made extensive use of bamboo in practice.

Our collection begins with discussions on the interaction between peoples and the environment in four distinct regions. Robert Marks shows how the Pearl River Delta, today's 'factory of the world', is itself an artifact, largely created over the past two millennia by the actions, purposive and incidental, of waves of in-migrants. This chapter presents an extraordinarily clear example of how human and natural systems have coevolved over a long and reasonably well-recorded historical period. Tan Yan and Lesley Potter then turn our focus to today's most famous, or notorious, reshaping of the human and natural environments—the resettlement processes of the Three Gorges Dam on the middle Yangtze River (Chang Jiang). Even more extensive, covering over two-thirds of the land area of the People's Republic of China and nearly three-tenths of its population, are the government's programs to promote sustainable economic development in western China. Wang Yiqian and Chen Yong, from Sichuan University, discuss the importance of focusing on ecologically fragile belts (EFBs) in policies for environmental conservation and ecosystem rehabilitation in this region. These are areas where natural hazards, poverty, and minority ethnicities tend to coincide. Cang Ming follows by making a strong case for promoting ecologically motivated migration into urban areas out of some of the karst and high, cold mountainous environments in this region that are least amenable to habitation.

Part Two focuses on the forestry sector. Liu Dachang provides an overview of the importance of policy both in stimulating three great nationwide waves of deforestation (the Great Leap Forward, the Cultural Revolution and decollectivization), and in promoting reforestation and preservation in recent years. Clearly, in this sector environmental degradation is not linear or due solely to demographic or economic factors. William Hyde and Jintao Xu reinforce and extend this point, noting the net growth in forest cover and standing volume of timber between 1978 and 1998, a period when the real economy per capita quadrupled. This restoration is all the more remarkable given the severe deforestation that took place in the early 1980s. The authors point to a number of rural environmental problems that are of greater real concern than deforestation, especially the untreated effluents from the pulp and paper industry. Abe Ken-ichi then sharpens the focus by investigating farmer-managed forests in minority areas of Yunnan Province, providing case study texture to the narrative. In particular, he notes that growth in plantation forests often occurs at the expense of biodiversity. Following on from this point, Yin Shaoting details the long history of one kind of plantation forest, rubber trees, in the Xishuangbanna area of Yunnan, and the growing recognition of its adverse effects on the natural and human ecologies.

Like trees, people need water. Part Three looks at the complex and dynamic interactions between people and water in different regions of China. Leiwen Jiang et al. show that in the huge but arid Tarim Basin in western Xinjiang Province, large-scale migration from eastern China, especially in the 1950s and 1960s, placed enormous pressure on the biodiversity and river ecology of the *tarim* ('convergence of waters' in the local Uygur language). Yet again, the problem has been as much one of unwise policies in cropping (especially cotton) and water pricing as of overpopulation or economic growth per se. Nakawo Masayoshi adds a long-term climactic perspective on the availability and use of water in the Tarim Basin and the second largest inland basin in China, the Hei in Gansu Province. He finds a dynamic interaction over time between wildly fluctuating populations and water regimes. James E. Nickum looks at China's 'marginal' capital, Beijing, which seems to be continually beset by the threat of a water crisis, but which has been able to find solutions—so far. Katri Mehtonen, in a chapter especially commissioned for this volume, looks at the Lancang/Mekong basin that bestrides another margin, the border that separates China from Southeast Asia. She indicates that the current focus on poverty reduction and environmental protection in Yunnan does not address transboundary impacts, in part because of their controversial nature.

Part Four provides three contributions that deal with the often endangered grasslands that cover large areas of China's landscape. Tsui Yenhu uses a detailed exploration of one specific township in the vast Altai grasslands of the northwestern corner of China to explain the serious degradation of grassland in Xinjiang over the past fifty years. As in the case of water, the problems appear to stem from non-sustainable policy orientations, particularly of local governments, that include ecologically inappropriate cropping choices, as much as from population or economic growth stresses. Tsui notes, however, that pursuing a genuine sustainable development strategy may require decisions that are too painful for real-world governments to actually make. Kobayashi Tatsuaki offers a historical perspective and a contemporary case study of desertification in the sandy lands of the Ordos Plateau bracketed by the great bend of the Huang He (Yellow River). Finally, Manuel Ruiz-Perez and his multinational colleagues take us back to southern China to provide a well-illustrated discussion of the socioeconomic and environmental roles of China's most famous grass, the 'poor man's timber', bamboo.

It is difficult to generalize from as rich and varied a collection of local studies as this volume presents. One of the points that comes out of this collection is that extreme caution is advised in generalizing about China's environment. Nonetheless, two things stand out in these papers. First, while recent changes in relations between people and the natural resource base in China have been dramatic and rapid, they have not always been out of line with the past. Second, where problems have occurred, policy failures have been at least as much at fault as growth in population or economies. These policy failures in turn are often part of a 'political ecology' (cf. Lye et al. 2003), where the ethnic, geographical and environmental peripheries are impacted by expansions of majority peoples and their habits from the core areas, and by decisions made in Beijing for non-local purposes such as nation building and economic development.

Notes

1. One reviewer objected to our implicit 'binary opposition' in juxtaposing 'human' and 'natural' (or for that matter, 'environmental'). We grant that the distinction between the two is not as clear. At the same time, it is analytically useful to consider the realms of human agency and all else as separate, coevolutionary systems that may or may not be in opposition.

References

Buck, Pearl (2004[1931]), *The Good Earth*, New York: Washington Square Press.

Lye, T., W. De Jong and K. Abe (eds) (2003), *The Political Ecology of Tropical Forests in Southeast Asia: Historical Perspectives*, Kyoto Area Studies on Asia, Volume 6, Kyoto: Kyoto University Press, and Melbourne: Trans Pacific Press.

Smil, Vaclav (1984), *The Bad Earth: Environmental Degradation in China*, Armonk: M.E. Sharpe.

1
Geography is not Destiny: Historical Contingency and the Making of the Pearl River Delta

Robert B. Marks

Introduction

At first glance, one might think that Figure 1.1 is an image of a beating heart surrounded by a dense network of arteries and veins. Although in truth this is a Landsat photo,[1] the heart metaphor is not far off the mark. The image depicts the Pearl River Delta and estuary in South China: Hong Kong is the peninsula at the lower right, the city of Guangzhou is just left of center, and Macao (not in the image) is across the bay from Hong Kong.

Today, thirty-five million people live in those ten thousand square kilometers, making it among the most densely inhabited places on Earth. Moreover, since the economic reforms of the early 1980s, the Pearl River Delta region has been the heart pumping China's emergence as a global economic power. As noted by the *Los Angeles Times*, 'if it was an independent country, the greater delta [area]...would be East Asia's fourth-largest economy and its second-largest exporter. Its sprawling, gray cities... collectively draw foreign investment at an astounding rate of nearly $2 billion a month' (August 24, 2003).

Indeed, the rapid industrialization of the Pearl River Delta region has put tremendous strains on the environment, with increasing signs of the collapse of some ecosystems. At dusk, the sun appears blood red through the haze of industrial smoke. For centuries a center of silk manufacturing, in the 1990s the mulberry trees and silk worms were replaced by fruit trees, not because there was greater demand for fruit, but because the air quality was so bad the silk worms died. Driving on two-lane roads at night, headlights rarely flash the eyes of any animals, and there is virtually no road-kill, a rough indicator of wildlife populations.

Given the speed of the industrialization and the extent of the environmental degradation now afflicting the Delta, environmental historians might suspect

Figure 1.1 Landsat photo of Pearl River Delta, December 24, 1975

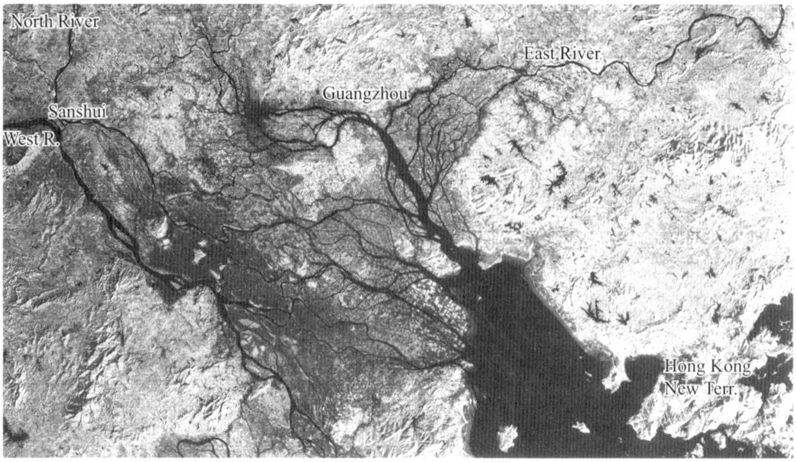

a familiar, declensionist storyline: industry and human activity despoil yet another subtropical forest region. But that is not what happened. Those who know a little Chinese history might suppose that the vast population increase of the eighteenth and nineteenth centuries led Chinese to deforest the Delta, leaving a vast farmland upon which later factories could be built. But that is not what happened either.

When I first began researching the environmental history of South China (see Marks 1998), I had assumed that the Pearl River Delta was a natural phenomenon that had been lying there awaiting human exploitation, in particular land clearance for settled Chinese agriculture. What I discovered instead is that the Pearl River Delta is very much a human creation.

Two thousand years ago, the Pearl River Delta did not exist. Chinese were just beginning to enter the region, and their small settlement of Nanhai (which ultimately became the city of Guangzhou) was not in a delta, but rather was a port overlooking an estuary dotted with islands. Indeed, for most of the first millennium CE, there was not much increase in the size of the alluvial delta. But starting around 1000 CE, and then accelerating rapidly during the fourteenth century, the Delta took shape, nearly reaching its current size (as in the Landsat image) by the early nineteenth century. What accounts for the timing and increased rate of creation of the Pearl River Delta?

The story is basically one of environmental history. Broadly understood, environmental history examines the mutual interactions of humans and their environment, and asks two related questions: in what ways does the environment condition human activities, and how do human actions change

Figure 1.2 China ca. 2000

the environment? In short, environmental history strives to put 'nature' back into the stories we tell about the past and about ourselves as humans. That is what I propose to do here with the Pearl River Delta, presumably a most natural phenomenon.

When historians usually think about 'the environment', most often it is 'natural geography' and the background, or stage, upon which human history is played out. Mostly the assumption is that the environment simply *is*—it is there, important in some ways (usually in terms of limiting what is or is not possible, especially economically), but mostly just lying there, unchanging. After an initial bow to 'geography', 'natural endowments', or 'the environment', most historians are happy to move on to the real stuff of history—the actions and interactions of humans.

I have to admit that my initial assumptions about the relations of people to the environment in South China included positioning humans as actors and the environment as the passive receptor of human actions. In my mind's eye, what I assumed to have happened goes something like this: for millennia, the natural processes of erosion and siltation gradually created the Pearl River Delta—a swampy, jungle-like place—long before Han Chinese pushed into the south, and probably even before aborigines settled there too. When Chinese arrived, first in small numbers during the Qin and Han, but then in greater numbers in later centuries, they used their military might and greater technological prowess to push the non-Chinese aside and to take the best farmland for themselves and the Chinese way of life. I imagined the Chinese coming into South China and then fanning out into the river valleys, including the Pearl River Delta, cutting trees and draining swamps, 'reclaiming' (*kai ken*) this fertile wasteland for productive agriculture, and then, as population density in the core increased, spreading out into the less fertile hill country.

Of course, my assumptions about how Chinese migrated into (and changed the environment of) South China had been informed by some very powerful images and theories about how Chinese history, in particular that of the late imperial period, was geographically structured, most of which we owe to G. William Skinner and his macro regional theories on China (see Skinner 1977a; 1977b; 1985; 1994). Using major watersheds and drainage basins as defining features, Skinner divided late imperial China into eight physiographic macro regions[2] within which, he argued, 'there developed' an urban system:

> We *start* (emphasis added) with the key fact that each region was characterized by the concentration in a central area of resources of all kinds—above all, in an agrarian society, arable land; but also, of course, population and capital investments—and by the thinning out of resources toward the periphery...[E]cological processes as well as technological (e.g. the transfer of fertility through erosion, on the one hand, and the use of irrigation and of fertilizer, on the other), boosted agricultural productivity in the lowland cores (Skinner 1977b: 216).

Implicitly, Skinner's formulation is a *history*, a story about how those macro regions developed over time. South China, the macro region Skinner dubs 'Lingnan' (more or less coincident with the provinces of Guangdong and Guangxi; see Marks 1998: 22–25), at first blush appears to fit Skinner's implied history. Three major rivers (the West, the North and the East) drain South China and converge at the headwaters of the Pearl River to form a delta (see Figures 1.1 and 1.3). Looking at contemporary maps, it is easy to imagine that that constituted the geographic structure of Lingnan when Chinese first began to settle there, leading ultimately to a process of development close to that outlined by Skinner: Chinese migrants settled the lower reaches of the river valleys and the Pearl River Delta, forming the 'core' of the Lingnan macro region, followed by later (and sparser) settlement of the hill regions

Figure 1.3 Lingnan ca. 1820

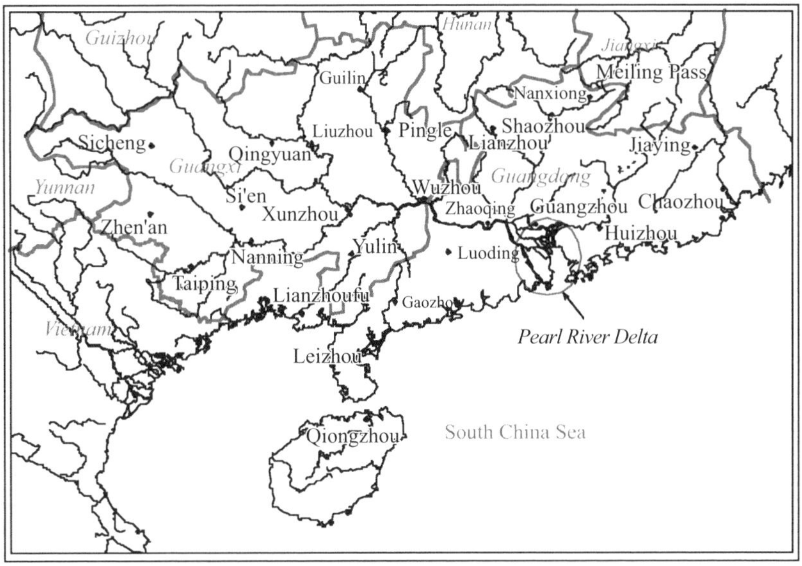

(Skinner 1997). This version of Chinese history sees 'the land' as a fixed thing just waiting for Chinese agriculturalists to 'open it up'. But that is not the way it happened, at least with regard to the Pearl River Delta.

I do not want to be misunderstood: Skinner's macro regional work is invaluable for understanding China, and I have used it to good effect in my work (Marks 1991; 1998). Fundamentally, I would argue that Skinner's macro regional theory is correct. But there are two assumptions implicit in his body of work that can lead to erroneous conceptions of China's historic development. First, his work on macro regions, based on nineteenth century data, assumes a particular physiography bequeathed to humans by nature and about which not much was or could be done. Second, he paints a picture of a macro region filling up with people, becoming densest first in the area that forms the 'core' and then 'thinning out' toward the geographic peripheries.

In a Skinnerian world of Chinese history, then, geography is destiny: core areas in river valleys and deltas are 'rich', while the periphery is 'poor'. And in Lingnan, what became the core was the Pearl River Delta. Contrary to the historical picture that Skinner's work implies, what I found when I examined the Pearl River Delta was quite different: geography was *not* tantamount to destiny. Rather, the existence of the Pearl River Delta and its development as the core of the Lingnan macro region was historically contingent. It is these historical contingencies that I now want to focus on.

Early settlement patterns and agricultural technology

Lingnan was first inhabited not by Chinese but by other peoples. To be sure, Chinese had settled in Lingnan as early as the creation of the first Chinese empire by the Qin (c. 221 BCE), albeit in small numbers, but even then they were invading territory inhabited for millennia by various non-Chinese ethnic groups. In the lowlands of the river valleys and along the coast were various peoples, the largest group now called the Zhuang, with a smaller group called the Li in the littoral belt on the Leizhou peninsula and the coastal strip on Hainan Island (Csete 1995). Over the centuries extensive interactions between Han Chinese conquerors and settlers and the Zhuang and the Li (from intermarriage to trade and learning the Chinese language, if not actually adopting Chinese cultural practices) rendered some of the aboriginal peoples less 'alien' than others. This gave rise to the distinction by the Song between the 'cooked' (*shu*) and the 'raw' (*sheng*) Li or Zhuang, the former referring to those who had accepted Chinese overlordship and the latter to those still beyond the influence of Chinese culture (Shepherd 1993: 7–8).

While the Zhuang and the Li were aboriginal in the sense of being the earliest known human inhabitants of Lingnan, other non-Chinese groups migrated into Lingnan in later periods and were seen by the Chinese as 'barbarians' (*man*). This characterization was later misunderstood by historians thinking the term meant 'aborigines'. The most important of these later immigrants into Lingnan were the Yao. Dating their arrival in Lingnan is controversial; some place it in the Han and others in the Yuan or even Ming dynasty (1368–1644). A recent Chinese assessment finds that the Yao first migrated in small numbers during the Han dynasty (202 BC–220 CE), but came from the north in larger numbers in the tenth century during the wars that gave birth to the Song dynasty (Mo 1985: 24). Whenever they first arrived, by the Ming the Yao had successfully colonized the uplands of northern Lingnan. In his mid-seventeenth century work on strategic geography, Gu Yanwu lists by name each of the mountains that the Yao had occupied: 106 in Qingyuan *xian*, or forty-one in Xinyi *xian*, for instance (Gu 1879: part 3: 11a–17b).

Whenever the Yao arrived in Lingnan, their preferred habitat was the hills and mountains where they organized themselves into villages, or at best into groups of villages that might be called tribes, and practiced slash-and-burn, shifting agriculture (Leong 1997: 31). In contrast, the lowland Zhuang peoples grew wet rice (Maglioni 1975: 23–24), cast bronze and wove silk. By about 300 BC a state called Yue had formed. With a state system and knowledge of sericulture and metallurgy, the Yue were the only people the Chinese did not consider barbarians. This did not prevent the Qin from conquering the Yue, an endeavor carried out over a period of twelve years (234–222 BCE). The Yue ruling elite fled south to organize another state in Thailand, leaving behind the Zhuang agriculturists in the lowlands.

How many indigenous people inhabited Lingnan before it was brought into the Chinese empire, and thus before Chinese began settling there, is not known. If the average density was as low as one person per square kilometer, then there may have been about half a million people in Lingnan at around 200 BCE. During the Han dynasty (202 BC–220 CE), a 2 CE count of households (adjusted for the boundaries of Lingnan) placed the total at 72,000 households, or 350–400,000 people, depending on assumptions about household size.[3] Citing earlier studies, Harold Wiens argued that the Han totals included both the few Chinese who had settled in Lingnan and the indigenous people, although the total omitted significant numbers of non-Chinese (Wiens 1967: 180–81). A guess at a half million is thus not unreasonable.

Predating (or concurrent with, in the case of the Yao) the Chinese settlement of Lingnan, then, the half-million or so non-Chinese peoples had developed two different agricultural regimes, one for the lowlands and one for the uplands (Moseley 1973: 12). In the lowlands, the Zhuang planted wet rice, maybe in paddies and maybe in the same plot year after year, while in the uplands the Yao slashed and burned their way through the hills, probably waiting twenty to twenty-five years before burning them once more, and on the coastal littoral and Hainan Island, Li peoples pursued a more desultory approach to food production.

Chinese migration into and settlement of Lingnan

Chinese migration into Lingnan came in three principal waves, the first a small one following the Qin subjugation of the Yue kingdom around 225 BCE when some hundred thousand or so troops occupied Lingnan and then intermarried with local Zhuang women. The second wave came in the early fourth century when nomadic tribesmen invaded north China and sacked the Chinese imperial capital at Loyang, bringing on the ‘Yongjia Panic’ when inhabitants of north China fled south. The third wave began similarly in the twelfth century when central Asians—this time the Jin armies (predecessors of the Mongols)—took the Song capital in Kaifeng in 1126 CE, forcing the Song to relocate their capital south of the Yangzi River in Hangzhou. This

Figure 1.4 Population of Lingnan, 2–1391 CE

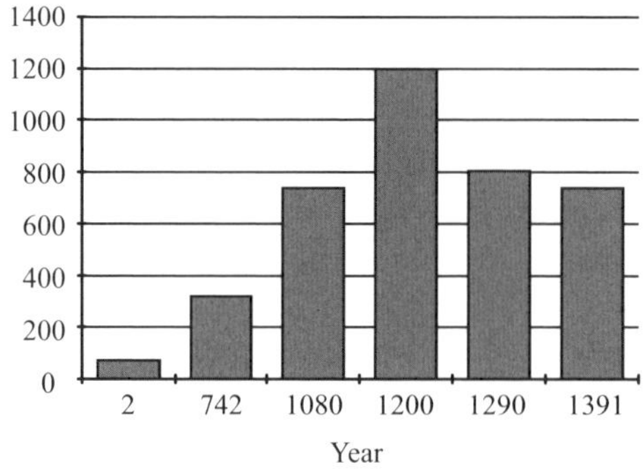

Sources: For 2 CE, *Han Shu*, Dili zhi, juan 8 xia; for 742 CE, *Xin Tang Shu*, juan 33; for 1080 CE *Yuanfeng Jiu Yu Zhi*, juan 9; for 1290 CE, *Yuan Shi*, juan 62–63; for 1391, Liang (1980: 277–78).

third wave continued into the 1270s when the Mongols began their push to conquer the rest of China (Li 1993: 171–205).[4] As many historians have noted, when northern nomadic invaders pressed south of the Great Wall they set off a wave of Chinese migration south.[5] Many of those fleeing the war and disorder in the north found their way through the passes of the Nanling mountains and into Lingnan.

Taking a *tour d' horizon* of Lingnan's population over the fourteen centuries from 2 to 1391 AD, we can see the impact both of the waves of in-migration and of war (see Figure 1.4). From the Han (2 AD) to the Tang (742 AD), the population had fluctuated higher and lower than the Han figure of 72,000 households, probably reaching a nadir late in the third century. From this point the population increased by the Tang (in the eighth century) to 319,000 households, largely as a result of fourth-century migration from the north. The population growth over the three centuries from the Tang to the Song (742 to 1080) was most likely the result of the rate of natural increase in the base population rather than in-migration. My estimate for the population in 1200 may well be low, given the massive influx of refugees after the 1126 loss of the Song capital in Kaifeng. Whatever the population may have been at the peak of the southern Song around 1200—and it certainly was higher than it had been in 1080—it declined substantially after the Mongols subdued South China in 1278, and again a century later during the establishment of the Ming dynasty.

By 1400, then, the population of Lingnan had gone through a great cycle, increasing more or less steadily from the Tang to the southern Song, and then declining for another two centuries until the beginning of the Ming. Relative to the population peak in the southern Song, Lingnan in 1400 had become relatively depopulated, reaching a nadir from which population increased slowly but steadily for another 250 years until the wars and destruction of the mid-seventeenth century Manchu conquest.

Moreover, the population was not evenly spread over the landscape. The highest population densities were not in the Pearl River Delta, as Skinnerian analysis would lead us to believe. Instead, until the Song dynasty (or possibly the Tang), the highest population densities were in the northern hill region of Guangdong province, in particular Nanxiong department. After the Mongol conquest, the highest population densities were on the Leizhou Peninsula and even on Hainan Island, not around Guangzhou. Why didn't those areas—Nanxiong in the north, or Leizhou in the south—become the core of the South China macro region Skinner called Lingnan? How and when did Guangzhou and the Pearl River Delta emerge as the population core of Lingnan? The answer to these questions links changes in the distribution of population in Lingnan to significant environmental change in what came to be the Pearl River Delta.

The making of the Pearl River Delta

For the first millennium after Chinese settlers moved into Lingnan, what we now call the Pearl River Delta and know as one of the most agriculturally rich and productive regions of China, second only to the Yangzi River delta, was not yet a delta, but in fact was still open sea, albeit a fairly shallow bay. Residents of Guangzhou, then called Nanhai, or 'the south sea', looked out onto a bay dotted with islands, probably appearing much the same way the South China sea looks today from Repulse Bay on the southern side of Hong Kong Island.

Certainly, silt carried downstream in the West, North, and East Rivers had been settling out in the bay, slowly creating the upper reaches of the delta; but because the silt content of these rivers was exceptionally low, the natural processes by which the delta was being created worked extremely slowly. Then, beginning in the eleventh century (during the Song dynasty), the delta began to grow more rapidly, and in the fourteenth century (during the Yuan dynasty) accelerated even faster. The series of four maps of the Pearl River Delta (see Figure 1.5), chosen to correspond with the dates for which we have population data, shows the formation of the delta. During the seven centuries from the Han to the Tang, the delta barely changed with the

Figure 1.5 The Pearl River Delta, 2–1820 CE

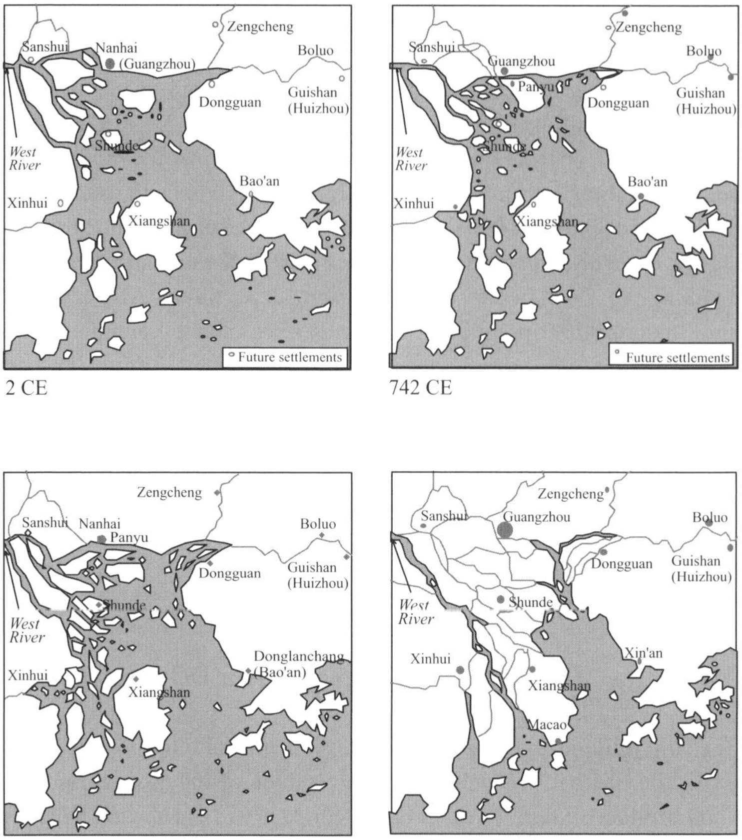

2 CE

742 CE

1290 CE

1820 CE

Source: Tan (1975–82).

bay remaining filled with water. By the Song dynasty, however, enough of the delta had emerged south of Guangzhou to block the view of the ocean, and by the Yuan dynasty alluvial sand bars appeared off the coast of Dongguan where the East River emptied into the bay. The largest increases in the size of the Pearl River Delta, however, occurred from the Yuan dynasty on. Where virtually no change had occurred in the nine centuries after the founding of the Han, in the 300 years from 1290 to 1582, what had been the island of Xiangshan became connected to the mainland (Zhou 1987: 58–69). Both the change in the shape of the delta over time and the rate of the change itself are interesting and significant: what accounts for these processes?

In the title to this section, I referred to the 'making' of the Pearl River Delta. I chose the word 'making' specifically to refer to the action of human beings, for more than anything else, people made the Pearl River Delta. Involved in this story are, in chronological order: the early settlement patterns and agricultural technology of the Chinese in-migrants; the building of water control projects in the lower reaches of the West, North, and East Rivers; the Mongol invasion of South China in the 1270s and the consequent displacement of the Chinese population from northern Guangdong to islands in the Pearl River estuary; and the creation of new lands off the islands in the estuary.

Slash-and-burn, the earliest and most rudimentary form of agriculture, was practiced in Lingnan until Chinese wet rice cultivation techniques spread to Lingnan in the eleventh century. When Chinese migrated into Lingnan first in Han times and later in greater numbers in the fourth century, they settled in the upland areas around Nanxiong and Guilin (see Figure 1.3), rather than near the swamps in the river valleys further south (the reasons for this will be discussed below). Burning off tracts of land, settlers broadcast millet, barley and wheat (and maybe rice) seeds onto the ash, and then harvested a crop for two or three years until the fertility of the soil was depleted. Abandoning the burned-off ground, the settlers then moved on, burning a new tract of land and beginning the process again. The fires must have been impressive, as this eighth-century poem describing the process in Lianzhou in northern Guangdong graphically illustrates:

> Wherever it may be, they like to burn off the fields,
> Round and round, creeping over the mountain's belly.
> When they bore the tortoise and get the "rain" trigram,
> Up the mountain they go and set fire to the prostrate trees.
> Startled muntjacs run, and then stare back;
> Flocks of pheasants make *i-auk* sounds.
> The red blaze forms sunset clouds far off,
> Light coals fly into the city walls.
> The wind draws it up to the high peaks,
> It licks and laps across the blue forest.
> The blue forest, seen afar, dissolves in a flurry,
> The red light sinks—then rises again.
> A radiant tarn brings forth an old *kau*-dragon;
> Exploding bamboos frighten the forest ghosts.
> In the color of night we see no mountain,

Just an orphan glow by the Starry Han [milky way]:
It is like a star, then like the moon,
Each after the other, until at daybreak the wind dies away.
Then first comes a light which beats on the stones,
Then follows a heat which glows up to heaven.
They drop their seeds among the warm ashes;
These, born by the "essential heat" [yang],
Burst into buds and shoots.
Verdant and vivid, after a single rain,
Spikes of trumpet vine come out like a cloud.
The snake men chant with folded hands;
Neither plowing nor hoeing involve their hearts.
From the first they have found the temper of this land,
Whose every inch holds an excess of "essential cold" [yin].
(Schafer 1967: 54–55)

Chinese settlers in the north at first adopted the slash-and-burn techniques of the non-Chinese hill peoples, the Yao and Miao, both of whom practiced various forms of slash-and-burn, shifting cultivation, along with hunting and gathering (Leong 1997: 34). Setting fire to forested hillsides during the dry winter months, groups of families then broadcast or drilled dry rice seed into the ash, and harvested crops for a few of years. They then moved on to another patch, leaving the depleted ground to return to forest, relocating their encampment as well. Remnants of this system were observed in the early twentieth century, with Yao planting seedlings and leaving the reforested area alone for twenty years after burning it off and planting it for two years (Fenzel 1929: 45). The reason for such a long 'fallow', as scientists later learned, is that hillsides burned off too regularly turned into grassland with dense root systems very difficult to break up and plant, unlike the soft, porous soils under a forest canopy. Even for agriculture, forest was better than grassland, and the Yao seem to have been aware of this fact.

The extent of the original forest burned off by settlers is not at all clear, nor is it clear whether the forest was given sufficient time to recover its composition of broadleaf evergreens before being burned off again, or whether the scrubby *pinus massoniana* took the place of the broadleafs. What does seem to be quite certain, though, is that the slash-and-burn agriculture practiced in the upper reaches of the drainage system opened the hills to greater erosion and hence to a higher silt content flowing downstream in the West, North, and East Rivers (Chen 1981: 243).

Until the eleventh century, much of this silt did not reach the Pearl River Delta, but rather was deposited in the flood plains in the lower reaches of the West, North, and East rivers. Much like the end of a garden hose under high pressure, the lower courses of these rivers flopped around from one outlet to another when monsoonal rains sent water gushing through the system, filling the flood plains with silt-laden water. When the flood waters receded, nutrient-laden silt was left behind. Clearly, the flood plains of the lower reaches of these rivers, especially the area near the confluence of the West and North Rivers, thereby contained fertile soils with great agricultural potential—but they had two related problems: flooding and malaria. Before the future Pearl River Delta could become the densely populated, agriculturally rich center of the Lingnan regional economy, the Chinese would have to either change the swampy environment of the flood plains of the West, North, and East Rivers, or else adapt to that environment, for the South China lowlands were inhospitable to northern Chinese people.

The role of malaria in slowing southward migration

To Chinese from the north, all of Lingnan looked diseased. In the words of the Tang-era writer Liu Xun: 'The mountains and rivers of Lingbiao [Lingnan] are twisted and jungly; the vapors concentrate and are not easily dispersed or diffused. Therefore there is much mist and fog to cause pestilence' (Liu 1975: shang 1). As is now known, of course, malaria is caused by a parasite transmitted to humans from a particular kind of mosquito, the *Anopheles*. Because of the linkages among parasite, mosquito and human host, malaria requires a specific set of environmental conditions to exist, and cannot spread beyond those limits. Malaria is thus not a communicable disease like plague, small pox, or cholera that can spread broadly and fast through human populations regardless of where they reside, but is limited to certain environments.

Tropical forests such as those originally covering Lingnan are especially good breeding grounds for parasites of all kinds. But the particular parasites that cause malaria in humans—several species of the single-celled protozoa belonging to the genus *Plasmodium*—were not present just waiting for human hosts to invade, but rather arrived with their human hosts when they migrated into South China. To be sure, malaria is found in monkeys, apes, rats, birds and reptiles, many of which inhabited Lingnan's tropical forests, but these forms are not infectious to humans. Human malaria is a very old disease, not merely evolving with humans, but even influencing the process of natural selection (McNeill 1976: 47–48). The disease probably originated in tropical Africa, and only later spread elsewhere with the Neolithic revolution. According to Bruce-Chwatt (1988: vol. 1: 3), the pre-eminent historian of the disease, in Neolithic times the infection established itself in Mesopotamia, India, the Nile Valley (from which it invaded the Mediterranean) and South China (see also Maegraith 1989: 201). Whether it was brought into Lingnan by the original settlers in the area, or was spread among an indigenous population by later arrivals, is not known. However and whenever malaria became established among the human population of South China, it was there among the indigenous valley-dwelling Zhuang population long before the first Chinese migrants arrived.

Anopheles mosquitoes carry three species of plasmodium harmful to humans, and each causes a different kind of malaria. Two cause intermittent fevers and are not particularly virulent, even in non-immune populations, but one—the malignant tertian, or pernicious malaria, caused by *p. falciparum*—is the most deadly to humans (Maegraith 1989: 201). Depending on the parasite, the clinical manifestations of malaria (including fever with or without paroxysm, sweating, and chills, vomiting and diarrhea, anemia, and hardening of the spleen) occur as the brood of the parasite undergo schizogony together (Maegraith 1989: 210–220). That all three forms of malaria were found in Lingnan is clear from the classification of fevers given by Zhou Qufei in the late twelfth century. In the 'lighter kind (*qingzhe*), the fevers come and go'. In the 'serious kind (*zhongzhe*), there is only fever and no chill'. And in the 'really serious kind (*geng zhongzhe*), fever continues without letting up' (Zhou 1178: juan 4:3b–4a). Zhou may have been erroneous in ranking the severity based solely on the periodicity of fever, but he clearly identifies three kinds of fever, all consistent with scientific understandings of malaria. Which of the three was predominant in Lingnan can only be guess work, but given what is known about the general epidemiology of the parasites and the fact that the malaria in Lingnan was often fatal to newcomers in the region, *p. falciparum* probably was the most prevalent.

With most of South China's rainfall coming in the summer when temperatures are in the range optimal for mosquitoes to breed, with pools of water forming in depressions in the earth's surface but especially in the swamps left by the annual flooding of the river systems, and with relatively high humidity, both the parasite and the *Anopheles* mosquito could flourish in Lingnan (Molineaux 1988 vol.2: 915). Since the parasite lived in both the mosquito and in humans, the environmental conditions which brought them into contact provided the environment for malaria. Human beings, as a host to the parasite, are thus a prerequisite for the existence of malaria: no humans, no malaria.[6]

That malaria was prevalent throughout much of Lingnan is certain, as Liu Xun's comment above makes clear. Some areas were known as being particularly virulent, such as around Qingyuan up the North River from Guangzhou. Other places mentioned malaria in the 'climate' section of their gazetteers, such as Chaozhou in eastern Guangdong where malaria was prevalent in the hills (Chaozhou fuzhi 1762: juan 2: 1b, 4b). But it was also apparent to Chinese observers that malaria was not found in all areas of Lingnan. In the Song dynasty toward the end of the twelfth century, for instance, Zhou Qufei could write: 'Not all deep[ly forested] and wild (*shen guang*) places in Ling[nan] are necessarily malarious. For example, Qiongzhou on Hainan Island and, on the north side of the sea [i.e. on the mainland across from Hainan] are described as *shen guang*, but there is little malaria'. But other *shen guang* places—Nanning, Yulin, Qinzhou, and Guiping (all in Guangxi)—all had it (Zhou 1178: juan 4: 2b–3a). Although malaria had spread to Hainan Island by the nineteenth century, even in the fifteenth it was still described as a place free of malaria. The littoral belt of southwestern Guangdong (or at least part of it)—described by Gu Yanwu (1879 juan 98: 22a) as an area with 'very little malaria'—was still free from malaria in the late nineteenth century. The area around Guilin seems to have been free of malaria: according to Fan Chengda, writing in the twelfth century, 'everywhere south [of Guilin] is the home of malaria' (quoted in Schafer 1967: 132). Northern Guangdong province, in particular Nanxiong, Lianzhou, and Shaozhou prefectures, was also free of malaria. Local ecological factors, such as elevation, played a critical role in accounting for the presence of malaria in one area or its absence in another, even if we cannot now be sure exactly what those ecological differences were.

The Chinese in-migrants to Lingnan understood neither the causes of malaria nor the environmental link to the mosquito, but they did have enough knowledge of where the disease was and where it wasn't to guide decisions about where to settle and where to remain. Most came to Lingnan via either the Meiling Pass or the Ling Qu canal, and then settled in northern Guangdong and Guangxi provinces. This was in part because these were the first regions 'south of the mountains' they encountered, and in part because they were free of malaria. Once settled, they tended to remain there, and the primary reason given was fear of malaria in other parts of Lingnan. According to the genealogies of several lineages that trace their roots to Nanxiong, during Song times fear of malaria kept them from migrating elsewhere in Lingnan, even in the face of mounting population pressure in northern Guangdong (Gu 1879: juan 98: 28a).

In the south part of Lingnan, significant Han population densities were found not in the Guangzhou region, but along the largely malaria-free southwestern littoral, including the Leizhou peninsula and the areas to its east and west. Guangzhou prefecture was not initially the most densely populated part of Lingnan, despite the fact that by the Ming dynasty it (in particular the Pearl River Delta) was to become the core of the Lingnan macro region. Certainly one possible explanation for the reticence of Chinese immigrants to settle the Pearl River Delta was malaria. In terms of its original physical environment, the flood plains in the lower reaches of the West, North, and East rivers seem to have been nearly ideal for the *Anopheles* mosquito. In other parts of the world where the mouths of rivers created deltas and swamps, malaria thrived, such as in the south of France and in much of Italy (Bruce-Chwatt 1988: 12–13; Scott 1939, vol. 1: 131–32).

Settlement patterns also provide indirect evidence that malaria existed in the flood plains in the lower reaches of the West and North Rivers.[7] Malaria cannot exist without human hosts to provide a 'reservoir of malaria infection' (Maegraith 1989: 201), so areas uninhabited by humans thus could not have been malarial. As noted earlier, an indigenous population had lived in the river valleys and along the coast, providing the infectious reservoir of humans that allowed malaria to become endemic: uninfected *Anopheles* mosquitoes picked up the parasite from an infected human and passed it along to a person uninfected or not recently infected. Interestingly, though, communities in which malaria is endemic also develop a certain immunity to the disease (Maegraith 1989: 204–05).

The indigenous population of Lingnan, especially those who lived in the lower reaches of the river valleys, would thus have acquired a certain level of immunity to malaria before Han Chinese even appeared on the scene. But for those without acquired immunity—such as Chinese migrants from north China—the disease would have been deadly. Besides malaria, other tropical diseases too no doubt ravaged newcomers from the north, posting warning signs for later arrivals. Indeed, the Yao (and later the Hakka) too settled in the hills and avoided the flood plains, perhaps as much because of their fear of tropical diseases as of their preference for the hills. If malaria was one reason Chinese (and Hakka and Yao) initially avoided the area around the Pearl River Delta, then to settle there Chinese either had to acquire immunity to malaria like the indigenous peoples, or change the ecological conditions so they were no longer so conducive to breeding *Anopheles* mosquitoes. The best piece of direct evidence for the gradual acquisition by some Chinese of immunity comes from a Ming-era author, Wang Linting, who claimed that 'in the Tang and Song, malaria in Lingnan was wherever [Chinese] in-migrants (*qian ren*) lived. But by the end of the Song, when the worthies and ministers fled [in the face of the Mongol invasion], [the earlier migrants] had become more like the locals [in not contracting malaria]' (quoted in Su 1982: 16).

Malaria by itself might have proved sufficient to keep Chinese out of the Pearl River Delta, but the fact was that the lower reaches of the West, North, and East Rivers were sufficiently swampy and difficult to farm whether or not they were malaria-infested. It is not that the river valleys were unpopulated, for the non-Chinese Zhuang people had inhabited those regions for millennia. Also, of course, there was the Chinese port city of Guangzhou, which, being set upon a hill overlooking the Pearl River estuary (Zeng 1986), was probably free of malaria, as may have been the area north of the city walls. Chinese settlement of the nearby malarial valleys, however, required levees to control flooding and to drain swamps, opening the flood plains to Chinese-style settled agriculture. Human beings were about to alter the environment to make it more hospitable to them, and inhospitable to the *Anopheles* mosquito as well (Tuan 1970: 129); ecological change cuts both ways.

How Song flood control projects prepared the way for settlement

Lingnan experienced two kinds of water control problems: too much water resulting in flooding, and too little or irregular supplies of water during the growing season. The summer monsoon rainfall pattern and the Lingnan drainage system accentuated both these issues. When the monsoons brought the rain to Lingnan, most fell in the four summer months, swelling the oft-times nearly dry river beds. From the west, all of the rain gathered into the catchment basin that emptied into the West River, and then spilled from Wuzhou down into Guangdong province. From the north, all of the rain gathered into the North River. The West and the North River joined at Sanshui ('Three Rivers') some ten miles west of the city of Guangzhou, forming the headwaters of the Pearl River. From the east, rain drained into the East River basin, pouring into the Pearl River to the east of Guangzhou.

The normal rainfall patterns thus poured huge amounts of water into the system in a very short period of time. Naturally, the flood plain from Zhaoqing (on the West River) down to Sanshui flooded every year, depositing ever greater amounts of silt eroded from the burned-off hills further up-river. As early as 809 (the Tang dynasty), levees constructed downstream from Zhaoqing prevented the West River from following the southern of two branches to the confluence with the North River, restricting the flow to the northern branch alone (Chen 1981 vol. 10: 183). The levee not only opened a flood plain to agriculture, but also sent all of the silt-laden flood waters further downstream, thereby increasing the pressure and flooding around

Figure 1.6 Levee construction on the West River, Tang–Ming dynasties

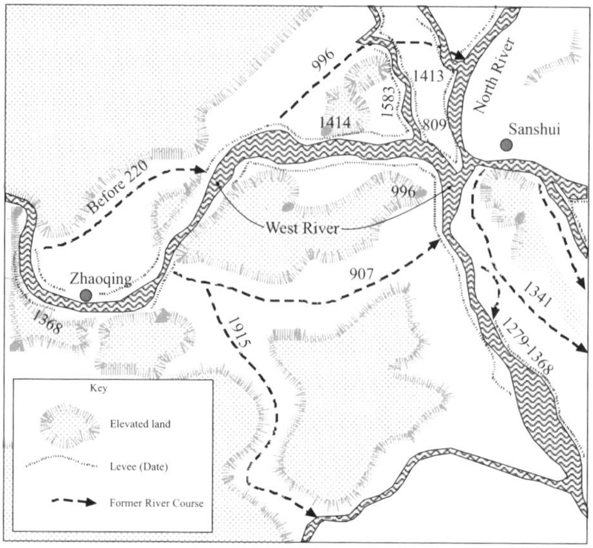

Sanshui (see Figure 1.6). Controlling the flooding at the confluence of the West and North Rivers thus was among the first large-scale water control projects in Lingnan to be tackled.

Around 1100, work commenced on the Sang Yuan Wei (or Mulberry Garden) enclosure. When it was completed, it was about twenty-eight miles in length and protected some 6500 *qing* (about 100,000 acres) of land from flooding (Wang 1986: 569–71), inaugurating a new era of agricultural development in that part of Lingnan. About the same time that the Sang Yuan Wei was being built, sea walls along the coast in Leizhou prefecture and in western Dongguan *xian* were also under construction. In Leizhou, nearly 25,000 *zhang* (about forty-eight miles) of sea wall was constructed during the Shaoxing era (1131–62 CE), creating some 10,000 *qing* (about 160,000 acres) of arable land by protecting it from periodic inundation by tides and typhoons, and no doubt eliminating the mangrove forests there as well (Guangdong tongzhi 1822). In Dongguan, the 12,806 *zhang* (about twenty-four miles) Xian Chao (Salt-Tide) dike was built in 1089, making 21,028 *qing* (about 300,000 acres) of land arable (Wang 1986: 581–82).

According to statistics compiled by Ye Xian'en and Tan Dihua, in Song times twenty-eight earthenwork dikes or embankments were built in the upper reaches of the Pearl River Delta, totaling 66,024 *zhang* (about 125 miles) in length and protecting 24,322 *qing* of land (nearly 400,000 acres); during the Yuan dynasty an additional thirty-four embankments were built, adding 50,526 *zhang* (about ninety-six miles) to the length and 2320 *qing* (Ye and Tan 1984: 73). In other words, the embankments extended for about 200 miles (or about 100 miles of river, if dikes were on both banks), and protected about twenty percent of the cultivated land in Yuan-era Guangdong. Additionally, in Chaozhou prefecture in eastern Guangdong, twenty-two sections of river dikes restrained the flood waters of the Han River in Haiyang and Jieyang *xian*, protecting another 88,000 *qing* of land (Guangdong tongzhi 1561: juan 26).

These flood control levees had the effect of directing each river into a single course, so that rather than meandering and spilling into numerous courses during the monsoon season, rivers ran straight for the bay. The flood plains, of course, were then opened for agricultural production. But these waterworks begun in the Song had other environmental consequences as well. Draining the swampy backwaters that had remained after the flood waters receded altered the ecological conditions that had favored the malaria-carrying *Anopheles* mosquito, rendering the areas so opened less deadly to Chinese originally from the north.

Equally importantly, though, the flood control works channeled the silt away from the former flood plains and directed it further downstream to the head waters of the Pearl River estuary. As a consequence both of the slash-and-burn agriculture in the hills and of the water control works, the amount of silt pouring into the Pearl River thus increased significantly from the eleventh century on. The changes to the Pearl River Delta that the increased amount of silt precipitated are evident in the maps of the delta. What is not evident, however, is that while more silt entered the upper reaches of the delta, it just might have continued to flow further into the bay had it not been 'captured', as will be explained below, by pioneers who fled to the rocky islands in the bay in the aftermath of the Mongol invasion of the 1270s.

Present at the creation: The Delta lineages from Zhujixiang

Many of the large, powerful families that came to dominate the economic, social, and political life of the Pearl River Delta by the Ming and Qing dynasties traced their origins to the Nanxiong area in northern Guangdong, in particular to the small but important village of Zhujixiang, which sat astride the road to the Meiling Pass (Li 1993: 183–92). All traffic flowing between Lingnan and the Yangzi River over the Meiling Pass—and that was a considerable amount of foot traffic, carrying products made from the natural wonders of Lingnan north and bringing the porcelains from

Jingdezhen south—had to pass through Zhujixiang. Indeed, so great was the amount of Chinese goods being transported south from the Yangzi River valley to Guangzhou for export to Southeast Asia, India, and the Arab world in the Tang period that one Zhang Jiuling had the pass 'chiseled' to provide a wider (and safer) path over the mountains, leading directly into Zhujixiang (Marks 1998: 21–27).

Perhaps the village had acquired its name—'Pearl Alley'—either from pearl merchants who may have operated there, or from the small, rounded stones inlaid into the ground to form the road. Qu Dajun claims the village was named after an especially filial Tang-era man from Nanxiong who was so fat he had 'a belly like a pearl' (Qu 1700: juan 2: 65). However the village obtained the 'pearl' part of its name, it literally straddled the road, with travelers forced to walk between the rows of houses and shops lining the 'alley'. Zhujixiang served as a resting place for merchants and other travelers before beginning or ending the one-day journey over the pass, and, we would assume, exercised some control over who passed through and how easily they did so. Nanxiong had been the first area in Lingnan settled by Han Chinese, and the population swelled after the Jin invasion of north China in the 1120s sent refugees fleeing south; many of those who took up residence in Zhujixiang, according to later chroniclers, were among the most wealthy and powerful families from north China. Resources and wealth thus concentrated in Zhujixiang, at least until the 1270s and the Mongol invasion.

Perhaps because the wealthy families of Zhujixiang had settled there in the first place out of fear of northern invaders, when Kublai Khan's armies began moving south in the 1270s, many residents decided to flee in 1273–74, a couple of years before the Mongol armies pushed south to the Nanling mountains (Xinhui xiang tuzhi: 84–85). When the Mongols crossed over the Meiling Pass in 1276, those residents of Zhujixiang who could flee did so; those who failed to flee probably died in the battles that devastated much of Nanxiong and Shaozhou. According to the traditions recorded in several genealogies, ninety-seven families (*jia*) with thirty-three different surnames (*xing*) fled south into the area that would become the Pearl River Delta (Chen 1982: 144–49).

Fearing the Mongols more than malaria or adversity, and perhaps hoping to find additional maritime escape routes, these families settled on the small islands that dotted the Pearl River estuary; many of the large lineages in Xinhui county, for instance, trace their origins to Zhujixiang (Xinhui xiang tuzhi: 85–87). Even today, the relics of that settlement pattern can be seen in the towns situated on what once were islands, but now are hills in an ocean of alluvial soil (see Figure 1.7). Most expressive of this is the town of Shawan, which hugs the southern slope of the island/hill where the town of Panyu was also built; to its west was the mouth of the North River (Liu 1992: 34–41; Liu 1995: 21–43). Shawan means 'bay of sand', which in 1276 no doubt it was.

How much cultivable land was available to the refugees on the islands in the Pearl River estuary in the late thirteenth century is not known. No doubt some alluvium had been deposited by natural processes, perhaps speeded up a bit by the slash-and-burn agriculture practiced in the hill regions around places like their former home in Zhujixiang. As the water from the West and North Rivers flowed into the bay and around the islands, the current slowed on the 'leeward' side, allowing the silt to settle out; but the new residents in the bay were not content to let natural processes create their agricultural land.

The creation of shatan (sand flat) fields

In what became the Pearl River Delta, settlers created new fields from the sand bars that formed wherever the current slowed sufficiently for the transported sediment to settle, but mostly on the downstream side of islands, or on the outward side of river bends. Called '*shatan*', or 'sand flats', these lands were truly new, having literally arisen from the water. Unlike polders or enclosed fields which had been reclaimed from swamps or coastal flats, the *shatan* 'grew' in the Pearl River, adding land where none had previously existed.

The particular topographical and hydrological conditions of the Pearl River estuary and the modifications to both caused by diking contributed to the creation of *shatan*. Before the Sang Yuan Wei and other dikes had been built at the headwaters of the delta, the flood waters of the West and North Rivers spilled over the river banks, depositing the sediment in the swamps bordering the river channels. Some of the silt was carried further down into what was then a bay, creating the delta. But when the dikes were built to prevent flooding, the river course was fixed and the sediment did not settle until further downstream. In Ming and Qing times, this occurred mostly in southern Panyu, northern Xiangshan, Shunde, eastern Xinhui and western Dongguan counties, and it is in these counties where *shatan* emerged.

Certainly, some of the *shatan* emerged by natural processes, but the majority were constructed. The method of creating *shatan* was relatively simple, but did require years until the land was usable. When a sandbar arose by natural means close to the water level, rocks were thrown around its perimeter not merely to fix the existing sand in place, but also to capture more sediment. After a more substantial enclosure was built, the sediment was 'transformed' by planting legumes (which fix nitrogen into the soil). After three to five years, the *shatan* would be ready for rice (Peng 1990: 164). According to the late-seventeenth century writer Qu Dajun, a three-year fallow period followed three years of growing rice (Qu 1700: juan 2: 57).

Once one *shatan* was created, more silt would build up on its downstream side. This silt would be captured by the process described above to create more *shatan*, and so on, until a whole series of *shatan* extended the cultivated land area to several tens of thousand *mu*. These connected *shatan* were called 'mother and child' (*mu zi*) *shatan*, rendering metaphorical the relationship between the original *shatan* and the one to which it had given birth. Continuing the Chinese metaphor, one could say that over time whole families or even lineages of *shatan* constituted the Pearl River Delta. More to the point, the delta had been built by people working with natural processes, but in the unusual conditions created by the Mongol invasion of South China.

Curiously, the abandonment of northern Guangdong farms in the face of the Mongol armies may have hastened the creation of *shatan* in the years or decades following. J. R. McNeill has shown that the soil of farms abandoned in the Mediterranean hill country in the aftermath of the fall of the Roman empire eroded rapidly, contributing both to the degradation of the mountains and to the deposition of silt in the lowlands at the mouths of rivers (McNeill 1992: 85, 191, 312, 319–22). A similar process may have occurred in Lingnan following the Mongol invasion, with increased erosion from abandoned northern Guangdong farms sending silt down the North River to the Pearl River. The soil from their farms may well have followed the refugees south to the Pearl River estuary where they recaptured it to farm once again. But an important difference with the Mediterranean experience should be noted. In the lowlands of the Mediterranean, the silt was not captured and turned into new land, but rather formed swamps and marshes, creating the conditions for malaria and rendering the lowlands uninhabitable until swamp drainage projects in the eighteenth and nineteenth centuries eliminated the disease. In the Pearl River Delta, by contrast, the new settlers captured the silt and turned it into *shatan*, perhaps preventing the creation of conditions favoring malaria in the first place.

One further aspect of the delta's creation is noteworthy. Because of the large amounts of labor required to create *shatan* (Yang 1959: 26), most projects were undertaken by wealthy families. All of the evidence comes from the Ming and Qing dynasties, but that should not obviate the point. Professor Liu Zhiwei (Liu 1992: 34–41) details the case of the He lineage in Panyu county, while Professors Ye Xian'en and Tan Dihua cite additional evidence about another He family, this one in Dongguan, which built or accumulated some 28,000 *mu* of *shatan* during the Kangxi, Yongzheng, and Qianlong reigns in the Qing dynasty (Ye and Tan 1985: 34–35; see also Peng 1990: 168). But not all *shatan* were created by wealthy lineages; Helen Siu (1989: 25) discovered a stele in Xinhui county making reference to military colonies established in the mid-Ming for the purpose of settlement and land reclamation in that part of the delta.

Conclusion

The Pearl River Delta was thus not created by purely natural processes, and had not been simply waiting for Chinese to migrate from the north and reclaim it for agriculture. Rather, the creation of the delta was the result of a complex causal chain. Chinese immigrants into Lingnan preferred to settle in the hills of northern Guangdong, fearful of the diseases in the river valleys further to the south. Their land clearing eventually increased the silt content of the rivers flowing south, but most of that alluvium never reached the bay, being deposited instead in the lower reaches of the North, East, and West river valleys. Only the construction of dikes and levees there in the Song directed the silt-laden waters into the upper reaches of the Pearl River estuary. Even then, the silt might have continued to flow further out into the bay had it not been captured by refugees from northern Lingnan who had fled from the Mongols to the islands in the bay. The creation of the Pearl River Delta and its later emergence as the densely populated, agriculturally rich core of Lingnan was thus a historically contingent, rather than naturally determined, outcome. One can only speculate as to whether or not the same pattern would have been followed had it not been for the 'historical accident' of the Mongol invasion. Whatever the outcome may have been, by the time the Mongols were driven from China and the Chinese peasant army leader Zhu Yuanzhang succeeded in establishing the Ming dynasty in 1368, the preconditions for the economic development of the Pearl River Delta had been created.

Notes

1. This image was captured on December 24, 1975 by the first Landsat satellite. Landsat 1 (originally known as ERTS 1, for Earth Resources Technology Satellite) was in service from July 23, 1972 to January 2, 1978. Using remote sensing equipment with a ground resolution of forty meters, Landsat 1 took over 300,000 images. For South China, which

receives about eighty inches of rain annually, many of those images are of clouds. Fortunately, the summer monsoon season gives way to clearer, drier weather in the winter, and hence to this image. For a brief history, see *www.earth.nasa.gov/history/landsat/landsat.html.*

2. Skinner recently increased the number of macro regions to nine, adding the 'Gan Yangzi'. See Skinner 1997.
3. The population figures are given as 'households' (*hu*) rather than 'people' (*ren*) or 'mouth' (*kou*) because the count of households is generally considered to have been more accurate than *ren* or *kou*. The purpose here in any event is to gain a perspective on the movements in population, a purpose served just as well by enumerating households as people. For those interested in population estimates, it is not unreasonable to assume each household to have been comprised of five–six people. For a discussion, see Robert Hartwell, (Dec. 1982: 369).
4. For a brief discussion, see Li Zhuanshi et al. eds. (1993: 171–205). Li et al. also claim a fourth wave of immigrants arrived in the late Ming, but cite no evidence to support the claim; I have not found any either, and so do not include it here.
5. The Russian historian L. N. Gumilev has suggested that the movements of central Asian nomads across the steppe were caused mainly by changing patterns of rainfall in central Asia: 'It is easy to understand what an enormous part such changes in steppe climate played in the history of the nomads of Eurasia. Livestock cannot live without grass, grass cannot grow without water, or nomads exist without livestock. Consequently, all these form a single system in which the key link is water' (Gumilev 1987: 23).
6. Although it seems a simple proposition that mosquitoes breed in warm, wet surroundings, and thus that malaria should exist nearly everywhere where those conditions prevail and people live, nature is not that simple. Indeed, the issue is so complex that scientists still cannot pinpoint the precise conditions which breed the *Anopheles* mosquito. Indeed, even the particular species of *Anopheles* responsible for malaria has been shown to be six 'sibling' subspecies, only two of which carry plasmodium and infect humans. The plasmodium-carrying 'siblings' are nearly indistinguishable from other *Anopheles maculipennis* except for laying their eggs separately upon the surface of water rather than in 'rafts'. Certainly differences in mosquito habitat—however minute—have favored one kind of reproductive strategy over the other. Variables such as whether the water is standing or moving, the amount of sunlight or shade, the presence of gases or minerals dissolved in the water, and the

salinity of the water, all affect whether or not the *Anopheles* mosquito will breed. Other factors include competition for habitat by other species of mosquitoes, such as the non-threatening *culex*, and predation by insects and fish. Those caveats aside, though, it seems that the *Anopheles* tends to favor stagnant water: small pools of water formed after rains, or the swamps left after floods. See Bates (1949: 302–330); Fiennes (1964: 77); Molineaux (1988 vol. 2: 916).

7. *Anopheles* mosquitoes were collected in the Pearl River Delta in the early twentieth century, but that is not proof that they had been there long before. See Faust (1929: 133–37).

References

Bates, M. (1949), 'Ecology of Anopheline mosquitos', in Mark Boyd (ed.), *Malariology*, Philadelphia.

Bruce-Chwatt, L. J. (1988), 'History of malaria from prehistory to eradication', in Walther H. Wernsdorfer and Sir Ian McGregor, *Malaria: Principles and Practice of Malariology*, Edinburgh: Churchill Livingstone.

Chen, B. (editor-in-chief) (1981), *Zhongguo ziran dili* (The Natural Geography of China) vol. 4, *Dibiao shui* (Surface water), Beijing: Kexue chuban she.

Chen, L. (1982), 'Zhujixiang shi shi (The True History of Zhujixiang)', [Guangdong] *Xueshu yanjiu*, 6.

Chang, K. (1986), *The Archeology of Ancient China*, 4th edition, New Haven: Yale University Press.

Chaozhou fuzhi (Gazette of Chaozhou Prefecture) (1762).

Csete, A. (1995), 'A Frontier Minority in the Chinese World: The Li People of Hainan Island from the Han through the High Qing', PhD thesis, State University of New York at Buffalo.

Faust, E. C. (1926), 'An inquiry into the prevalence of malaria in China', *The China Medical Journal*, xl(10): 938–56.

Faust, E. C. (1929), 'Mosquitoes in China and their potential relationship to human disease', *The Journal of Tropical Medicine and Hygiene*, xxxii (10):133–37.

Fenzel, G. (1929), 'On the natural conditions affecting the introduction of forestry as a branch of rural economy in the province of Kwangtung, especially in North Kwangtung', *Lingnan Science Journal*, 7.

Fiennes, R. (1964), *Man, Nature and Disease*, London: Weidenfeld and Nicholson.

Gu, Y. (1879), *Tianxia junguo li bing shu* (The strategic advantages and disadvantages of all counties in the world), Shanghai: Shangwu yinshu guan.

Guangdong tongzhi (Gazette of Guangdong Province) (1561; 1822).

Gumilev, L. N. (1987), *Searches for the Imaginary Kingdom of Prester John* (translated by R. E. F. Smith), New York: Cambridge University Press.

Hartwell, R. (1982), 'Demographic, political, and social transformations of China, 750–1550', *Harvard Journal of Asiatic Studies*, 42:2.

Leong, S.-T. (1997), *Migration and Ethnicity in Chinese History: Hakkas Pengmin, and Their Neighbors*, Stanford: Stanford University Press.

Li, Z. et al. (eds) (1993), *Lingnan wenhua* (Lingnan culture), Shaoguan: Guangdong renmin chuban she.

Liang, F. (1980), *Zhongguo lidai hukou tiandi tianfu tongji* (Historical Population, Land, and Tax Statistics), Shanghai: Renmin Chuban She.

Liu, X. (1975), *Ling biao lu yi* (Strange records from Lingbiao) [Lingnan], Reprinted in *Qin ding si ku quan shu*, 138, Taibei: Shangwu yinshu guan.

Liu, Z. (1992), 'Zongzu yu shatian kaifa—Panyu Shawan He zu de ge'an yanjiu (The Beginnings of Lineages and *Shatian*—Studies of the He Lineage of Shawan, Panyu County)', *Zhongguo nongsh*, 4: 34–41.

Liu, Z. (1995), 'Lineages on the sands: The case of Shawan', in D. Faure and H. F. Siu (eds), *Down to Earth: The Territorial Bond in South China*, Stanford: Stanford University Press.

Los Angeles Times (2003), 'Supercharging China's engine', August 24: A4.

Maegraith, B. (1989), *Adams and Maegraith: Clinical Tropical Diseases*, Oxford: Blackwell Scientific Publications.

Maglioni, R. (1975), *Archeological Discovery in Eastern Kwangtung: The Major Writings of Fr. Rafael Maglioni (1891–1953)*, Hong Kong: Hong Kong Archeological Society.

Marks, R. B. (1991), 'Rice prices, food supply, and market structure in eighteenth-century South China', *Late Imperial China*, 12(2): 64–116.

Marks, R. B. (1998), *Tigers, Rice, Silk and Silt: Environment and Economy in Late Imperial South China*, Cambridge: Cambridge University Press.

McNeill, J. R. (1992), *The Mountains of the Mediterranean World: An Environmental History*, New York: Cambridge University Press.

McNeill, W. H. (1976), *Plagues and Peoples*, New York: Anchor Press/ Doubleday.

Mo, N. (ed.) (1985), *Guangxi nonye jingji shi gao* (Draft History of the Agricultural Economy in Guangxi), Nanning: Guangxi minzu chuban she.

Molineaux, L. (1988), 'The epidemiology of human malaria as an explanation of its distribution, including some implications for its control', in Walther H. Wernsdorfer and Sir Ian McGregor, *Malaria: Principles and Practice of Malariology*, Edinburgh: Churchill Livingstone.

Moseley, G. (1973), *The Consolidation of the South China Frontier*, Berkeley: University of California Press.

Peng, Y. (1990), *Qing dai tudi kaiken shi* (The History of Qing-era Land Reclamation), Beijing: Nongye chuban she.

Qu, D. (1974[1700]), *Guangdong xin yu* (New Sayings from Guangdong), Hong Kong: Zhonghua Shuju.

Schafer, E. H. (1967), *The Vermilion Bird*, Berkeley: University of California Press.

Scott, H. H. (1939), *A History of Tropical Medicine*, Baltimore: Willliams and Wilkins.

Shepherd, J. R. (1993), *Statecraft and Political Economy on the Taiwan Frontier, 1600–1800*, Stanford: Stanford University Press.

Siu, H. (1989), *Agents and Victims in South China: Accomplices in Rural Revolution*, New Haven: Yale University Press.

Skinner, G. W. (1977a), 'Cities and the hierarchy of local systems', in G. W. Skinner (ed.), *The City in Late Imperial China*, Stanford: Stanford University Press.

Skinner, G. W. (1977b), 'Regional urbanization in nineteenth-century China', in G. W. Skinner (ed.), *The City in Late Imperial China*, Stanford: Stanford University Press.

Skinner, G. W. (1985), 'Presidential address: The structure of Chinese history', *Journal of Asian Studies*, xliv(2).

Skinner, G. W. (1994), 'Differential development in Lingnan', in T. P. Lyons and V. Nee (eds), *The Economic Transformation of South China: Reform and Development in the Post-Mao Era*, Ithaca: Cornell East Asia Series.

Skinner, G. W. (1997), 'Introduction and Map 1.1', in Sow-Theng Leong, *Migration and Ethnicity in Chinese History: Hakkas Pengmin, and Their Neighbors*, Stanford: Stanford University Press.

Su, G. (1982), 'Song dai de Guangxi shehui jingji' (The society and economy of Song-era Guangxi), *Guangxi shifan xueyuan xuebao*, 4: 75–84, reprinted in *Zhongguo renmin daxue shubao ziliao she fuyin basokan ziliao, jingji shi*.

Tan, Q., editor in chief (1975–82), *Zhongguo lishi ditu ji* (Historical Maps of China), Shanghai: Ditu chubanshe.

Tuan, Y. (1970), *China*, London: Longman Group Ltd.

Wang, P. (1986), 'Qing ji zhu jiang sanjiaozhou de nongtian shui li' (Irrigation of Agricultural Fields in the Pearl River Delta during the Qing), *Jindai Zhongguo qugan shi yantao huilun wenji*, Taibei: Academica Sinica Institute of Modern History.

Wiens, H. J. (1967), *Han Chinese Expansion in South China*, The Shoe String Press.

Xinhui xiang tuzhi (Xinhui County Gazette) (1970), Hong Kong: Guangzhou Xuehui.

Yang, C. K. (1959), *Chinese Communist Society: The Family and the Village (A Chinese Village in Early Communist Transition)*, Cambridge: MIT Press.

Ye, X. and D. Tan (1984), 'Ming Qing Zhu Jiang sanjiaozhou nongye shangyehua yu xushi de fazhan' (The Development of Agricultural Commercialization and Markets in the Pearl River Delta during the Ming and Qing), *Guangdong shehui kexue*, 2.

Ye, X. and D. Tan (1985), 'Lun Zhujiang sanjiaozou de zu tian' (On Lineage Lands in the Pearl River Delta), in *Ming Qing Guangdong shehui jingji xingtai yanjiu*, Guangzhou: Guangdong renmin chuban she.

Zeng, Z. (1986), 'Cong lishi dimaoxue kan Guangzhou cheng fazhan wenti' (Looking at the question of the development of Guangzhou city from the perspective of historical physiography, *Lishi dili*, 4: 28–41.

Zhou, Q. (1178), *Ling wai dai ta* (Explanations of things beyond the passes), reprinted in *Qin ding si ku quan shu*, 138–139, Taibei: Shangwu yinshu guan.

Zhou, Y. (1987), 'Zhujiang sanjiaozhou de chenglu guocheng' (The formation of the Pearl River Delta), *Lishi dili*, 5: 58–69.

2
Three Gorges Project: Impacts of Rural Resettlement and Urban Relocation on the Environment in the Reservoir Area

Yan Tan and Lesley Potter

Introduction

With its varied climate, abundant hydropower, extensive natural resources and long history of human civilization, the Yangtze River basin has a vital economic and strategic position in China's development. The upper reaches of the Yangtze River, spanning 99°30'–111°25' East and 24° 30'–35°45' North, encompass a region from the source of the Yangtze (Tuotuo River) to Yichang city, Hubei Province. This region is of significant strategic importance to the Yangtze River basin and even to the sustainable development of the whole nation. However, environmental degradation in the Yangtze basin has become increasingly acute in recent years. Soil erosion is the number one environmental problem affecting the sustainability of social and economic development not only in the upper reaches but in the entire river basin and the country (Da 2001). The sediment content of the river water has increased greatly. Downstream courses have become severely silted up and flood control capacity reduced (Wang and Cheng 2000). Natural disasters have frequently occurred in the central and lower reaches, and in turn, affected the nation's economic development and improvement in the livelihoods of the people (Li and Luo 2001). The Three Gorges dam is situated at the lower section of the upper reaches of the Yangtze. The principal purposes of the Three Gorges Project (TGP) are to reduce the flood hazard downstream of the dam, provide electricity to meet the industrial and domestic demands in Eastern and Central China, and improve river navigation (CWRC 1997).

Human resettlement and land inundation are important environmental problems. Human displacement produced by dam construction is characterized by compulsory mass removal. Relocation and resettlement associated with the TGP include four components: 1) human displacement and reset-

Photo 2.1. Wanzhou, a major city to be inundated by the Three Gorges Project

tlement; 2) relocation and reconstruction of the cities and townships to be inundated; 3) removal of industrial enterprises affected by the project; and 4) rehabilitation of basic infrastructure (such as transportation and telecommunication facilities and equipment).

Due to a shortage of cultivated land, a difficult physical and economic environment and an underdeveloped infrastructure, the challenge to successfully resettle all the rural residents in the reservoir area is huge. The earlier policy on TGP resettlement simply stressed the reclamation of land and the removal of people up-slope within their origin counties. About sixty percent of the rural residents were designated to continue with agricultural activities (REG 1988). It was not until 1998 that the central government realized that deforestation in the upper reaches of the Yangtze River was one of the important factors responsible for serious floods and that clearing these areas for displaced people must be limited. This realization has led to significant policy changes placing greater emphasis on distant resettlement and closing down small enterprises which generate severe pollution. Resettling rural people via the 'near resettlement' scheme and subsequent policy changes have environmental implications for the reservoir area.

There are a number of positive impacts of rural resettlement and urban relocation. Capital investment makes it possible for: 1) socioeconomic development and environmental rehabilitation; 2) comprehensive development of agriculture, moving farming from sloping to terraced land and subsequently reducing water and soil erosion; 3) reconstructing city and town systems and improving urban environments; 4) enhancing the capacity of protective engineering works along the bank of the reservoir; 5) accelerating urbanization so that some rural residents can be resettled in urban areas, relieving the pressure of rural resettlement and reducing environmental pressure; and 6) inputting specific funds for environmental protection and rehabilitation.

The environmental problems resulting from the processes of rural resettlement and urban relocation are secondary consequences of the TGP. Some crucial issues have been given less attention than others, and this study attempts to make a contribution to this field. Section one briefly summarizes the environmental conditions in the study area. Section two examines and assesses the environmental problems in the processes of rural resettlement and urban relocation. Section three takes Kaixian County as a case study, revealing the contradiction between land lost to the reservoir and the demand for land for resettling rural migrants as a core environmental problem, causing a decreased human carrying capacity in the reservoir area. In section four some effective environmental measures are noted, such as farmland protection engineering works, reforestation and the development of eco-agriculture. Section five analyses the recent 'two adjustments' to resettlement policy to enhance environmental sustainability in the reservoir area. The 'two preventions' to forestall geological disasters and water pollution are also mentioned. Environmental policy implications for rural and urban resettlement are addressed in the conclusion.

Study area

The reservoir is canyon-shaped. At the normal pool level of 175m at the completion of the dam in 2009, the reservoir will stretch 660 km long and average 1.1 km in width, covering a total area of 1084 km^2. The reservoir will flood twenty cities and counties, 24,500 ha of cultivated land and orchards, and about 35 km^2 residential districts and 824 km of roads. The affected area also includes 108 historical and cultural monuments. Some 1.3 million people will be resettled and rural resettlers account for 42.7% of the total over a period of seventeen years (CWRC 1997).

The reservoir area is defined as the region that will be affected by the reservoir. Due to the vast size of the project and also for ease of management, the reservoir area is divided into two regions—Chongqing and Hubei—on the basis of flooding status, the numbers of people to be displaced, responsibilities and administration of displacement and resettlement (Figure 2.1). The Chongqing reservoir region is an area with the largest proportion of farmland to be inundated and population to be displaced. It makes up some eighty percent of the flooding losses (mainly farmland, urban sites, industrial enterprises, infrastructure and cultural relics) and resettlers to be

Figure 2.1 The Three Gorges Reservoir area

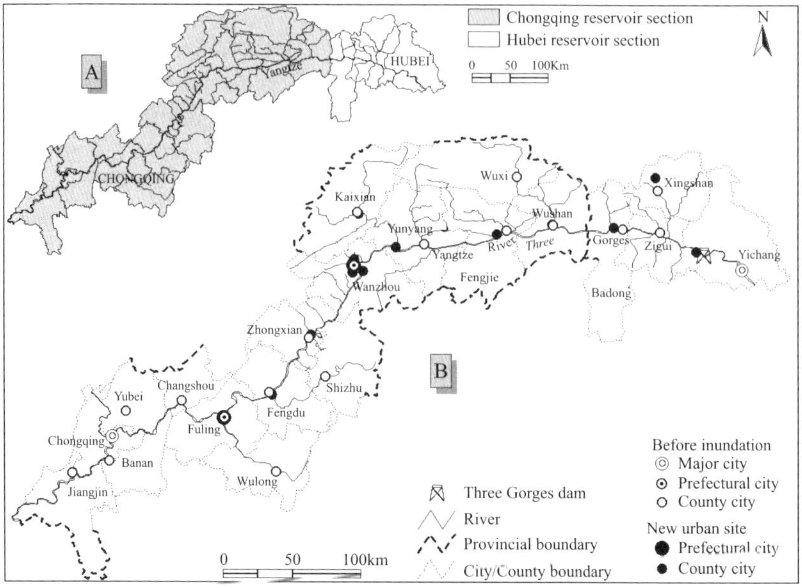

Note: 'A' shows the two sections constituting the Three Gorges Reservoir area: Chongqing and Hubei. 'B' illustrates the various cities and counties in the reservoir area, some of which will be completely submerged and others affected by the construction of the Three Gorges dam.

displaced. Chongqing is the main study area from which the case studies in the present paper are drawn.

Physical conditions of the reservoir area

The Three Gorges reservoir area has a complex and rugged topography, with mountains and hills occupying a large proportion of the total land area. Over seventy-four percent of the landscape is mountainous and 21.7% features low hills, while valleys and small plains make up less than 4.3%. The entire reservoir area has a typical sub-tropical monsoonal climate. The mean temperature in January is two to four degrees higher than in the middle reaches and downstream areas of the Yangtze. The reservoir area has an annual rainfall of some 1100 to 1200mm and a frost-free period of between 300 and 340 days (Chi 1996). Large areas of yellow alluvial soil are found in the valley, enabling it to be not only a double-cropping rice production area but also a center for planting citrus. The reservoir area has been the most favorable zone of orange production in China. There are more than one million *mu* (one ha = fifteen *mu*) of orange orchards in the reservoir area and orange production has become the fundamental economic source for farmers in some regions.

Underdeveloped economy

Due to its special relationship to the dam project, the area had received a small amount of construction investment from the central government prior to officially implementing the TGP in 1993 (Zhu 1996). The reservoir area is one of the eighteen continually poverty-stricken regions nationwide. Nearly ninety percent of the poverty-stricken population in Chongqing resides in the reservoir area. Per capita social gross production in the affected twenty cities and counties in the reservoir area was only about 45.2% of the national average in 2002, or 56.4% of the per capita national average GDP.

Fragile ecological environment

Arable farmland in the reservoir area is scarce. The cultivated land per capita is about one *mu* in the submergence area (Liang 1995). Sixty percent of the existing cultivated land is under arid farming, one-quarter to one-half of which has a slope gradient exceeding twenty-five degrees (Wang 1993). Due to a long development history, no large plot of gentle slopes is available in the reservoir area. With a total area of 54,200 km^2, the population density reached 314 persons/km^2, much higher than the national average (121 persons/km^2) in 1993. The areas above the 175m inundation line (the highest water level of the reservoir when it is completely filled in 2009) were more populous, having a population density ranging from 600 to 1000 persons/km^2. The human carrying capacity of the land in the reservoir area has been exceeded (Chen 1987). Present forest cover in this area is low. The counties from Fuling to Wanzhou only have a forest cover of between five to ten percent. Soil erosion is extremely serious throughout the whole region (Du, Shi and Yuan 1994). As the reservoir area has a weak tectonic structure prone to small earthquakes, it suffers from landslides and mud-flows, dominantly distributed in the section of the main channel from Yunyang to Zigui. The total population threatened directly by landslides and slumps exceeds 100,000 people.

Environmental impact assessment of rural resettlement and urban relocation

Institutions of environmental protection

The environmental impact assessment (EIA) of the TGP considered the entire Yangtze river basin as an integral system and divided it into several sub-systems: the reservoir area, middle and lower reaches, and river estuary region. The main findings of the EIA concluded that the TGP will exert both positive and negative impacts on the environment and ecological systems

(Research Institute for Protection of Yangtze Water Resources 1995). While positive impacts will occur mainly in the middle and lower reaches of the Yangtze, the negative impacts will be concentrated in the reservoir area.

In 1994, the Chinese government issued the *21st Agenda of China: Population, Environment, and Development White Book*. This agenda is a guideline in formulating the intermediate and long-term social and economic development of the country. To ensure appropriate growth and sustainable development of the national economy, the following strategic measures are proposed: 1) control of population growth; 2) energy savings and an increase in the efficiency of energy utilization; and 3) development and utilization of energy resources that cause little or no pollution.

In December 1993 the Environmental and Cultural Relics Protection Committee under the TGP Construction Commission (TGPCC) was established. It is to be in charge of environmental issues arising from the project. Meanwhile, the Environmental Protection Division was established for the purpose of environmental management of the dam construction site. 'The Ecosystem and Environment Supervision and Monitoring Plan of TGP' and 'The Environmental Protection Action Plan in the TGP Construction Site' were formulated in 1994 and have been implemented since then. The TGPCC established an Eco-environmental Coordination Group in July 1995. It consists of the relevant agencies and agents: the State Environmental Protection Administration (SEPA), the Chinese Academy of Sciences (CAS), the Three Gorges Development Corporation, the Resettlement Development Bureau under TGPCC, the Water Resources Ministry, the Agricultural Bureau, the Forestry Department, the Health Department, the National Meteorological Bureau and the leaders of the provincial governments in Sichuan (now Chongqing Municipality) and Hubei. This group is responsible for leading, inspecting and coordinating the environmental prevention work in the dam site, reservoir area, and mid- and down-stream regions of the Yangtze. A 'Bulletin on Ecological and Environmental Monitoring Results of the TGP' has been produced every year by the SEPA since 1997 (Wang 2000).

The main principles and evaluation criteria for China's environmental management, for instance, are the 'three spontaneities (*san tongshi*)', referring to activities in production, planning and pollution control: 'whoever develops, protects; whoever destroys, restores; and whoever uses, reimburses'. These principles are applied to urban relocation associated with the TGP. Compensation for environmental protection in the reservoir area has been explored (Huan 1996). Environmental protection relating to relocation in the reservoir area is being undertaken by the governments of Chongqing municipality and Hubei province. A proportion of the resettlement compensation package, 82.26 million *yuan* is allocated to an environmental protection fund—67.3 million *yuan* for Chongqing and 14.96 million *yuan* for Hubei (Fang and Chen 2002).

In the 'Constitutional Law of the People's Republic of China (PRC)', Article 26 notes that the State protects and improves the human and ecological environment, prevents and controls pollution and other public hazards. In the 'Environmental Protection Law of the PRC', issued on 26 December 1989, Article 19 specifies that measures must be taken to protect the ecological environment while natural resources are being tapped. Since 1989, China has implemented an 'Ecological Prevention Plan in the Upper Reaches of the Yangtze' (Wang 2000). The 'National Eco-environmental Construction Planning' has been carried out since 1999 (Liu 1999). The Three Gorges reservoir area is an ecologically fragile region where the ecology and environment must be protected and rehabilitated. The environmental impacts of displacement and resettlement have drawn much attention since flooding disasters occurred on the Yangtze in 1998.

Environmental consequences of resettlement

The main and direct environmental issues of the TGP are identified as the following: 1) environmental protection in the construction site of the dam; 2) environmental impacts of inundation; 3) impacts on public health; 4) impacts on cultural relics and the natural landscape; 5) impacts on water quality and temperature; and 6) impacts on aquatic life forms, the Yangtze estuary and coastal marine areas (Fang and Chen 2002). Human resettlement and land loss are a direct environmental consequence of the inundation of land and assets.

Considerable research has been conducted into the environmental impacts of the project itself (e.g., Chen, Xu and Du 1995). However, environmental problems and protection in resettlement were not addressed until 1999, when the Resettlement Development Bureau of the TGPCC issued the 'Measures for Administrations of Resettlement Construction Projects with TGP'. Thus no adequate assessment had been made to measure the potential impacts of the displacement activities on the reservoir environment (Li 2000: 291–292). The potential environmental effects of resettlement can be identified as the following (see Figure 2.2):

1. Land inundation and resettlement to nearby locations significantly increase the pressure on population carrying capacity of the reservoir area.
2. Land reclamation on hillsides to accommodate resettlement to nearby locations destroys the vegetation, exacerbating soil erosion.

3. Human displacement affects the demographic patterns, societal structures, cultural traditions and relationships of people.
4. The development of small- and medium-sized enterprises in urban areas as well as township-owned enterprises in rural areas can result in serious pollution, especially water pollution.
5. Urban relocation and engineering construction at the dam site may eventually improve the environment and benefit economic development after the final completion of urban reconstruction and the dam structure. However, it can aggravate water and soil erosion, pollution and vegetation destruction at the dam site and the resettlement locations. Much agricultural land has been forced out of production.
6. Geological disasters such as landslides and slumps might be caused by the reservoir when it is filled. Urban and infrastructure reconstruction and rural resettlement and re-building can worsen such hazards.

In addition, incidences of local epidemic diseases may increase due to the changes in the human environment brought about by the near resettlement and increased interaction of people in the reservoir area (Jiang 1999). The impact on people's health may become a serious problem in some areas, for example, in Kaixian where the county seat will be surrounded by relatively stagnant water when the reservoir is filled in 2009. The sludge collected at the bottom of the reservoir during the months of higher water level (October to May) will be exposed to the hot and humid weather during the months of low water levels (June to September), and will most likely lead to the increase of water-borne diseases. Currently there is yet to be an effective system in place to prevent and manage the diseases, which are closely related to the seasonal flooding and ebbing of the reservoir.

Environmental impacts of rural resettlement

The two biggest and irreversible environmental effects of the TGP are considerable inundation of farmland and displacement of some 1.3 million people. In the process of resettling the rural migrants, some adverse effects of land reclamation on steep slopes have occurred in the reservoir area.

Inadequate human carrying capacity of the land

The extent and degree of environmental impacts on the resettlement communities and regions will depend on a set of variables. The main variables include the general levels of economic development; land availability; geographical location; and spatial patterns of resettlement in the recipient locations.

Figure 2.2 Environmental Impact Assessment of the TGP Resettlement

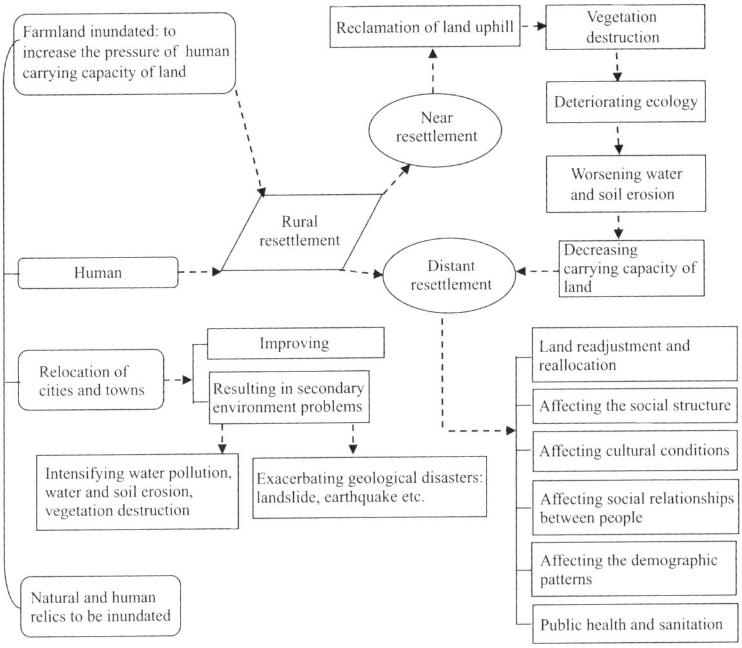

Due to the complexity and significant differences of these variables in different resettlement locations in the reservoir as well as distant resettlement communities, the consequences of environmental impacts may vary greatly. This study argues that: 1) 'near resettlement' mainly results in problems relating to the physical environment; and 2) the effects of 'distant resettlement' lie mainly in the human environment, including demographic, social and cultural aspects. The most conspicuous impact on migrants and host people is readjustment and redistribution of land among host villagers and migrants in a resettlement community. The environmental impacts of the distant resettlement schemes will perhaps need more time to manifest themselves. Attention to impacts on the migrants' health and on the resettlement communities is required.

Water and soil erosion

Deterioration of water and soil erosion is identified as another serious problem of the TGP (Du and Yan 1999: 299–303). A wide range of studies has been conducted on the sedimentation of the reservoir and on water quality (Wang 2000; Chen, Xu and Du 1995; Du, Shi and Yuan 1994). One study (Lu and Higgitt 2000) found a mean annual soil loss in the reservoir area during the last four decades estimated at 4500 t/km^2 per year. The extent of eroded land exceeds sixty percent of the total terrain in the reservoir area (Du, Shi and Yuan 1994: 5). Research has shown that for every one percent reduction in forest cover in the Three Gorges region, soil erosion increases by 128.47 t/km^2 per annum (GRICAS 2000). The amount of soil eroded from slopes makes up 60.6% of the total erosion, equivalent to 53,333 ha of fertile topsoil, with an annual loss of chemical nutrient elements of 1.2 million tons. Erosion rates are strongly related to slope gradient. Statistical data on 127 projects in the reservoir area in 1996 demonstrated that induced water and soil erosion have affected an area of 35.1 km^2, destroyed vegetation and land cover over 36.9 km^2 and discharged solid waste of 100 million m^3, most of which was poured directly into the rivers.

Human influence is a major factor resulting in serious soil erosion in the reservoir area. Owing to the far-reaching and profound impacts exerted by rural resettlement and urban relocation, much pressure is now being exerted on the fragile soil resources. The population size in 1988 was more than twice that of 1949. Increasing human demands for grain crops, firewood and timber usually results in deforestation. In the early 1950s, forest cover in the reservoir area was around thirty to fifty percent, but recently it has decreased to around ten percent, or even five to seven percent along the banks of the Yangtze. The ratio of cultivated land area to the total land area of a region has reached thirty-four percent, higher than the national average of ten percent. The proportion of farmland on slopes makes up seventy-five percent of the total, a quarter of which is located on slopes of more than twenty-five degrees. Sloping land with a gradient of more than twenty-five degrees is tilled. Due to long-term human activities including overuse and inappropriate development, soil erosion has become a serious environmental issue. Soil erosion occurs over more than ninety percent of the total Three Gorges area, with 45.5% of the soils being heavily eroded with an erosion coefficient of greater than 4000 t/km^2 per annum (Shi and Yang 1989).

In the reservoir area, there are 266,667 ha of steep land on slopes of more than twenty-five degrees, accounting for 18.6% of the total area. The proportion of steep land in the lower reservoir section amounts to 28.5% in Fengjie, 38.4% in Yunyang, 45.9% in Wushan and 49.6% in Zigui. A slope gradient of approximately ten degrees is an obvious limit. Over this threshold, the amount of soil erosion increases by 390 t/km^2 per annum for every increase of one degree. By contrast, the increment of soil erosion on forested land is only 39.3 t/km^2 per annum (GRICAS 2000). This also illustrates that improving steep land by creating terraces is necessary to restore the environment in the reservoir area.

Soils usually suffer from nutrient depletion, texture coarsening, thinning width and soil drought (Zhang 1996; Du, Shi and Yuan 1994; Yang 1994; Xu and Liu 1993). For some soils on the slopes, the rate of erosion exceeds the rate of formation and materials in these soils may be worn out, making the topsoil layer thinner and thinner. In the reservoir area, most soils are between thirty and fifty cm thick. Soil fertility is therefore very poor. Water and soil erosion reduce the depth, fertility and water-holding capacity of the soil, leading to a higher frequency of floods and droughts, landslides, mud and stone flows and other mountain disasters. It also reduces the life expectancy of the reservoir through sedimentation and causes water contamination (Du and Yan 1999).

Relationships between environmental and socioeconomic impacts

The construction of the TGP and associated resettlement will greatly affect the environmental, economic and social conditions in the reservoir area and other affected regions. All the factors resulting in the socioeconomic and environmental consequences of resettlement interact with one another. Hence, impact assessments in the three dimensions (economic, social and environmental) should be seen as related components of the general assessment of the resettlement, rather than as independent segments. The relationships between the three fields of impact assessments—economic impact assessment, environmental impact assessment, and social impact assessment—interact as shown in Figure 2.3.

Given the substantial spatial and temporal scale of the TGP resettlement, particularly distant resettlement, some effects have not yet emerged. Thus data to be collected and used in the impact assessments are very limited. However, considerations and attempts to seek out suitable and effective methods for complex assessments of resettlement are of higher significance than the assessment itself.

Land reclamation and low-yield farmland: A case study in Kaixian

Kaixian is the largest county to be flooded by the TGP in terms of the number of residents to be relocated and the area of land to be inundated. It is located in a densely populated mountainous region along the valleys of the Xiao River, one of the tributaries of the Yangtze. In the planned local resettlement communities, the total exploitable wasteland is 2285.5 ha, with 205.5 ha designated for cropping and 828 ha for forestry. Of all the land-use

Figure 2.3 Relationships between economic, environmental and social assessment of the TGP Resettlement

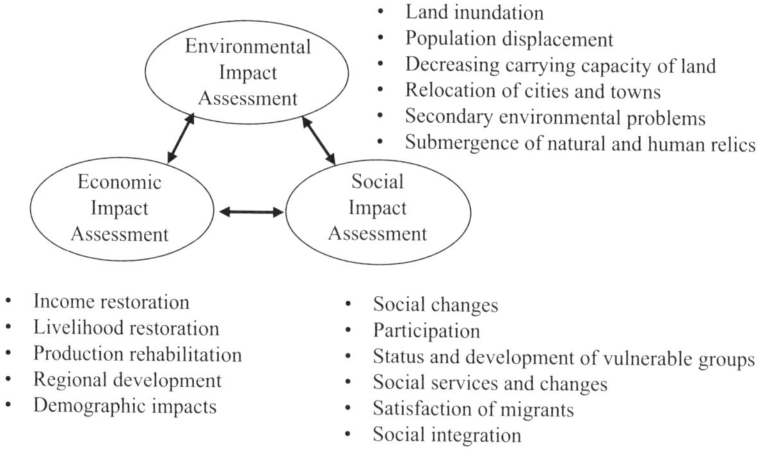

types in the receiving communities, dry land situated on slopes of more than twenty-five degrees makes up 46.8% (or 388 ha) of the total cultivated land (CWRC 1996: 30). Plans were made to reclaim land and improve low yielding fields over an area of 12,420.2 ha, involving 23.72 million *yuan*. Since 1986, when the 'trial of the TGP resettlement' was initiated, reclaiming land and improving poorly yielding farmland have been carried out in sixty-two villages within eleven towns in this county. By the end of the year 2000, 659.4 ha of land had been developed, of which 237 ha could be used to receive rural migrants. However, the portion which could be directly allocated to resettle migrants was only 68.2 ha because most of the reclaimed and improved lands are of rather inferior quality. Other important reasons are outlined in the discussion which follows.

Inadequate investment in land

Per capita funds for the 'productive resettlement population' (people whose land will be inundated and who need to be provided with new farmland or conditions to restore production) are only 7000 *yuan* in Kaixian (rate as in 1993). The gross compensation and funds for rural resettlement are 540.17 million *yuan*, or fifteen percent of the total resettlement funds in Kaixian. However, investment in land reclamation usually needs around 300,000 *yuan*/ha, while 120,000–150,000 *yuan*/ha is required to improve land on hillsides into standard terraced land with affiliated irrigation. Relying only on the resettlement funds, which actually are the main means of investment to reclaim or improve land, is likely to yield few results.

Photo 2.2. Kaixian county seat and its peripheral districts (to be inundated)

Inadequate water conservancy measures

The newly developed land suffers from heavy erosion. Human management of the land is not efficacious due to the long duration of resettlement from its commencement in 1993 to its completion in 2009. The newly developed land is situated on steep hillsides with thin topsoil and sparse vegetation. The lack of agricultural equipment and facilities (especially water conservation) to sustain the achievements of land development is another factor. Although the rainfall in this county is 1100–1500 mm/yr, due to inadequate water storage facilities, supply for human and livestock consumption and agricultural activities is still problematic. The capacity to resist natural disasters, particularly droughts and floods, is weak.

Steep dry land

More than ninety percent of the land to be reclaimed is rocky or sandy. Most dry land is distributed on steep hillsides (>twenty-five degrees), lacking in irrigation to cope with droughts. The soil layer ranges from ten to thirty cm in depth. Soil salinity is serious. The yield of agricultural products is low and annual agricultural income averages only around 4500 *yuan*/ha. A few years of farming are needed for newly opened land to become productive.

There is inadequate land available for resettlement and whatever land exists is often unsuitable for agriculture. Moreover, rehabilitating the uphill land of more than twenty-five degrees to forest or grassland will exacerbate the situation of limited arable land supply. By early 2001, only 1786 farmers had been resettled in the agricultural sectors in Kaixian. Persons settled near the town or county urban settings have only obtained around 0.4 *mu* of farmland per capita. Those settled away from townships have received about 0.6 *mu* per capita. Some 30,000 rural migrants are required to be resettled locally

Figure 2.4 Stages of displacing migrants in Kaixian (1993–2009)

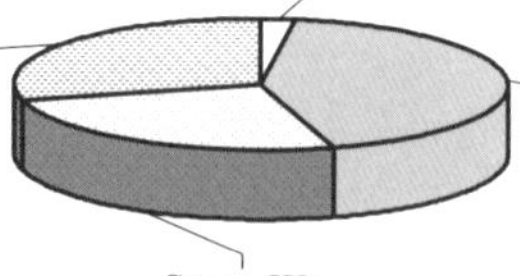

Source: CWRC and CSPDI (1996).

through the 'near resettlement' scheme. The main resettlement task will be concentrated in the latter stages (2004–08) of the resettlement, as shown in Figure 2.4. These migrants have to share the remaining unproductive land uphill with the host people.

Measures to protect and improve the environment

Practical countermeasures to tackle the issues of land shortage and other environmental concerns in the reservoir area include: 1) protecting farmland from inundation by adopting appropriate protection engineering works; 2) developing eco-agriculture; 3) improving low yielding cropland; 4) reforestation; and 5) relocating a proportion of rural residents through a distant resettlement approach.

Land protection works: A case in Kaixian

Benefits

The flooding areas in Kaixian are densely populated and mainly agricultural districts, which extend along the three tributaries (the Nan River, the Dong River and the Puli River) of the Xiao River. The fifteen land plots to be protected are located at the end of a backwater of the Xiao River (Figure 2.5). The fifteen zones encompass 392 villager groups of forty-three villages in seven towns. The waters will be shallow when the reservoir is filled. The average annual net income in these zones is usually twenty percent higher than the average income in the county. The land protection project

Photo 2.3. Vegetable growing area in Shuidong village, Fengle town, Kaixian (to be inundated)

aims at protecting the current cultivated land, orchard, infrastructure, and other assets from being submerged, enhancing the standard of local flood prevention, and minimizing the number of migrants.

The area of cultivated land and orchard to be protected is 908.3 ha. Parts of Fengle, Houba and Zhaojia town seats could be conserved. The current standard of flood prevention is only capable of controlling two-year floods (those that occur on average once every two years). The prevention works will be designed to control five-year floods for ordinary land and ten to twenty year floods in areas with high potential impacts on people's livelihoods or critical infrastructure (CWRC 1997). As such, the loss through inundation could be reduced by eighteen million *yuan* in urban areas and four million *yuan* in rural areas during a flood year. It is estimated that 26,034 rural residents may not need to be displaced, of whom 19,309 people are 'productive resettlement population' (KG and CSPI 2000). Correspondingly, the demand of population resettlement for land could be reduced. Some 1,773 ha of cultivated land and orchard might be relinquished.

Uncertainty of the protection project

The feasibility evaluation of the protection project should adopt an accumulative assessment, taking into consideration the interaction of economic, geo-engineering and environmental impacts on this region. There are few successful examples of land protection works in China's hydro projects. Over two decades of the TGP feasibility assessment, how to deal with the problems of land inundation and population resettlement that Kaixian encounters has been in heated dispute among experts and accountable institutions (CWRC 1997). The corresponding scale of

Figure 2.5 Fifteen land plots to be protected in Kaixian

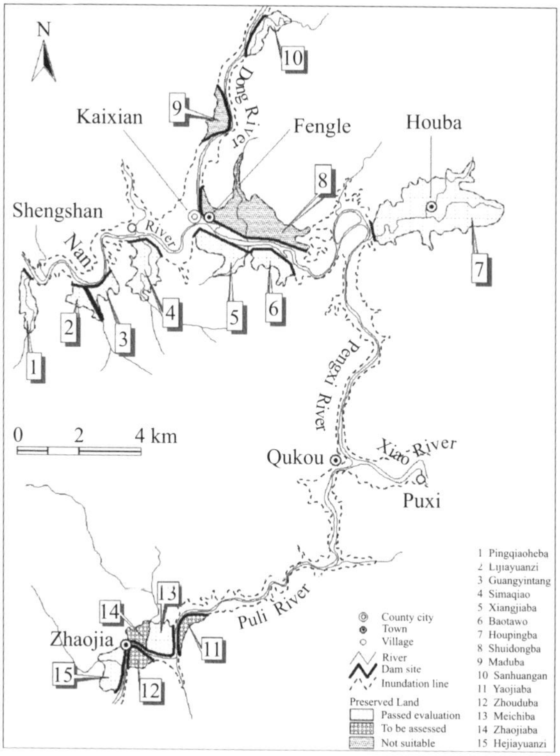

Source: Compiled from Kaixian Resettlement Bureau: Diagram of Land Protection Works (1993).

land protection measures along the Xiao River was one of the factors in determining the final height of the Three Gorges dam. The Changjiang (Yangtze) Water Resource Commission (CWRC) has conducted research into the feasibility and operational schemes of this protection project. The main schemes include: 'large-scale protection' which was rejected in 1996; and 'small-scale protection', which is more favorable from the cost-profit point of view.

However, the feasibility assessment and detailed plans for most plots of protection zones have been through long periods of planning but a conclusion has not been reached. The uncertain situation prevails in all the affected parts despite the fact that in 1995 the CWRC explicitly authorized implementation of a 'small-scale protection' project, starting with a feasibility study of each zone. In turn, this long interim period has increased the complexity and uncertainty for the local communities to implement any resettlement

Photo 2.4. Fertile farmland along the Xiao River in Qukou, Kaixian (to be inundated)

schemes. Resistance is met when authorities require rural people who reside in some zones to move to distant provinces. The survey in Fengle town found that most residents refused to be moved out, hoping to stay on their land which they believed could possibly be protected.

Developing eco-agriculture

The development of eco-agriculture is determined by the fragility of the environment and scarcity of land in the reservoir area. It is impossible for people to gain food security only on the basis of expanding farmland area and crop farming. Through improving production conditions, the transformation of agriculture from cropping to eco-agriculture is gradually realized.

One measure is to improve the capability of water and fertility preservation through building terraced land and equipping facilities. Terraced systems, especially terraced agro-forestry, can enhance the retention of organic matter and this was reflected in the microbial indicators. Taking into account economic costs and environmental influences, one measure is to develop newly opened farmlands and low yielding farmlands into high quality terraced land. Some trial resettlement villages in Fuling district of the Chongqing reservoir area have successfully carried out this strategy (Luo and Xia 2002).

Alternative land use systems, and especially agro-forestry, are currently being promoted in the reservoir area to enhance soil retention and improve soil quality. Studies have found that fertility of different soil types in the reservoir area has been improved, such as their total organic, chemical and nutrient status, soil microbial biomass, and other indices for measuring soil's physical and bio-chemical characteristics (Yan, Yang and Campbell, 2003).

Some trial districts of eco-agricultural development appear in Wanzhou, Kaixian, Yunyang and Fuling.

Another measure is to build 'household ecological economy' (*ting yuan jingji*), through integrating people's livelihood and production with ecology. In doing so, rural residents can utilize clean and renewable energy by developing methane generating pits using waste manure and grass. It is estimated that a ten m^2 methane tank is equivalent to one *mu* of vegetation used as firewood (Luo and Xia 2002).

Reforestation and its impacts

Cropping on hillsides is a main factor that exacerbates soil erosion, land degradation and environmental deterioration. However, agrarian production on dry land has played a key role in the rural economy in the reservoir area. Fifty-six percent of food grains are produced on sloping land. The current policy of returning the cultivated land on slopes of twenty-five degrees or greater to forest or grassland will benefit the rehabilitation of the environment in the reservoir area in the long term. However, it will reduce grain production, resulting in food insecurity (Zhang and Xu 1997). The implementation of the reforestation policy will greatly increase the pressure on rural resettlement in the reservoir area because available and exploitable arable land is so limited.

In March 2000 the Forestry Bureau, the State Planning Commission and the State Financial Ministry issued the 'Circular on Carrying out Preparatory Work of Returning Cultivated Land to Forest or Grass in the Upper Reaches of the Yangtze River and in the Upper and Middle Reaches of the Yellow River' (government document No. 111 [2000]). To achieve this goal, China adopted a policy of returning farmland to forests and pasture, closing off mountains to logging, planting trees, substituting grain for poverty aid, and signing contracts with individuals. The State committed itself to compensate farmer households for their loss of cultivated land. However, the subsidy standards in cash and kind are low. The cash subsidy is usually inadequate to cover the real cost of purchasing seedlings or grass seed, thus affecting the initiative of farmers. Another problem is 'over reforestation', in which the area of cultivated land returned to forest or grassland surpasses the intended figure. Thus, the affected households could not receive the corresponding grain and cash subsidy from the county or local governments. This usually results in social problems and affects the basic maintenance of farmers after reforestation.

To prevent new soil erosion, ecological rehabilitation in the reservoir region is urgently required. Soil erosion directly relates to sedimentation of the reservoir and affects the life span of the dam (Lu 1996). In compliance with the 'Eco-Environmental Construction Plan in Chongqing Municipality'

Photo 2.5. Cropping on steep slopes, common in reservoir area (Shuanglong, Wushan County)

(Liu 1999), some 133,300 ha of farmland for cropping, which is about eight times the area inundated by the reservoir, would be reduced if fifty percent of the steep land were restored to forest or grassland within ten years. Exploiting the near resettlement scheme through opening land on steep slopes is obviously in conflict with the reforestation policy.

Moreover, reforestation will reduce the amount of farmland in the area, especially in areas with concentrated slopes. The people who no longer cultivate their farmland can be re-allocated other land mainly through adjusting farmland from the host people in the same village or emigrating to local or distant resettlement communities. As a result, environmental emigrants will be produced. One piece of research (Wang, Yang and Guan 2001) indicates that the latent number of emigrants caused by reforestation will be huge. It is estimated that if the area removed from cultivation reaches forty percent of the currently reclaimed farmland, this number will be 890,000, which is equivalent to the number of re-settlers around the reservoir caused by the TGP. This should be considered as a potentially serious problem in implementing the reforestation policy in the reservoir area.

'Two adjustments' and 'two preventions'

The Chinese government has carried out a 'development-oriented resettlement' policy in the TGP displacement and resettlement project. The principal goal of this policy requires that involuntarily displaced persons be assisted in their efforts to improve their former production levels, income earning capacity and living standards, or at least to restore them to before-project or before actual displacement levels. The government's emphasis had been on the developmental resettlement policy mainly from an economic perspective before the 'two adjustments' to the resettlement policy were put forward in May 1999 (Zhu 1999).

'Two adjustments' and contributions to environmental rehabilitation

'Two adjustments'

One adjustment, commencing in 2000, is a shift from a policy of settling rural migrants on uphill sites in the reservoir area via 'near resettlement', settling rural migrants in 'secondary and tertiary industry sectors', and settling rural migrants by themselves through 'self-employed resettlement' to the 'three combinations' of resettlement approaches. The 'three combinations' are: between local resettlement and distant resettlement, between centralized resettlement and scattered resettlement, and between government-organized resettlement and voluntarily self-employed resettlement, based on a principle of resettling migrants in line with local conditions. The government has encouraged more rural migrants to move to more distant resettlement communities beyond the reservoir area. The other adjustment is a shift from simply restoring industrial enterprises in the reservoir area to their original scale, standards and functions to restructuring, improving the quality of the product and enhancing benefits through merging or shutting down unprofitable small or medium-sized enterprises. The structures of industrial organization, ownership and product must be readjusted in the process of relocation. Those State- or collective-owned enterprises generating serious pollution, without markets for their products, and with liabilities outstripping assets must be bankrupted or shut down.

Contributions of the environmental rehabilitation

Distant resettlement

In the land-based rural resettlement projects, the land supply in the resettlement communities is a pivotal determinant in the evaluation of the number of resettlers that it can receive. Moving some rural residents out of the reservoir area can prevent people from developing land on steep slopes and thus reduce soil erosion. 'Moving people out' lays the foundations for sustainable development in the reservoir area and realization of the target of 'becoming rich gradually' for rural resettlement.

Land scarcity is particularly serious in five counties—Wushan, Fengjie, Yunyang, Kaixian, and Zongxian—situated in the lower reservoir section of Chongqing. The rural migrants to be displaced and needing to be provided with land there amounted to 223,000 persons and the farmland to be inundated is 14,333 ha. Yet the land available for resettlement (through reclamation of waste land, shifting sloping land to terraced land, taking some land from the host people, and land protection works) is only 8,867 ha so that at least one third of the people displaced cannot be resettled locally due to land deficit. A total of 330 million *yuan* has been invested in reclaiming uncultivated land since the early 1990s. As a result, 10,000 ha of newly developed farmland has been provided; but more than half of the newly developed land is located on slopes of greater than twenty-five degrees (Wu 1999). If calculated on the basis of per capita 0.96 *mu* of land-holding before displacement, about 5,466.7 ha of arable land and orchards would be lacking.

Through the implementation of the 'Government-organized Distant Resettlement' (GODR) scheme in 2000–04, approximately 100,000 rural residents were displaced and resettled in the eleven designated provinces (Sichuan, Jiangsu, Shandong, Zhejiang, Guangdong, Jiangxi, Anhui, Shanghai, Hunan, Hubei and Fujian). The strategy of distant resettlement helps rural people avoid deforestation and cropping on steep slopes, thus reducing the possibility of water and soil erosion caused by rural resettlement. Distant resettlement has laid a solid foundation for sustainable development in the reservoir area.

Closing down small industrial enterprises

There are 1599 industrial and mining enterprises to be relocated in the overall reservoir area. Ninety-eight percent of these enterprises are small and unprofitable, mainly consisting of establishments manufacturing cement, fertilizer, brick, paper, wines and printing. According to the amended plan of Hubei and Chongqing, 1012 enterprises are about to be bankrupted. The remaining 587 enterprises will be reorganized into 406. In 2001 the relevant department issued a 'Circular of Issues on Bankruptcy of the State-owned Industrial Enterprises Below the 135m Flooding Line in the Reservoir Area', initiating the implementation of the industrial restructuring policy (Fang and Chen 2002).

Polluted water and solid waste are the main pollutants discharged by the industrial enterprises. Due to seriously inadequate facilities and equipment for sewage disposal, about 500 million tons of sewage are discharged into the Yangtze and tributaries annually (Fang and Chen 2002). The existing three 'pollution belts' in the Yangtze adjacent to the built-up areas of Chongqing,

Wanzhou and Fuling have become worse. Some new 'pollution belts' in rivers close to some county seats have emerged (Tao and Liu 2002). The large number of enterprises to be bankrupted will greatly reduce the pollutant sources. The remaining one third of enterprises cannot discharge liquid or solid waste until they are satisfactorily treated and meet emission standards. Water quality in the reservoir and in the middle and lower reaches of the Yangtze is expected to maintain its current level or even be improved.

'Two preventions'

Prevention of geological disasters

There are 2490 landslide locations in the reservoir region, some of which will be submerged by the reservoir. Among the other locations, some may be managed through engineering works and others only require close monitoring. People living on the 553 remaining landslide locations must be relocated (Tao and Liu 2002). Urban re-construction aggravates the original pollution and geological disasters such as landslides and mud-flows (Gu and Huang 1999). For instance, a landslide occurred in the old county seat of Yunyang on 17 January 2001 (*Guangzhou ribao*, 19 January 2001). An earthquake with a magnitude of 4.1 occurred forty km away upstream of the Three Gorges dam on 13 December 2001. They caused serious damage to the navigation on this section of the river and enormous economic losses. Senior Chinese water engineers foresee that the reservoir is likely to induce potential landslides at about 760 locations (Yang 2002), while new landslides may revive the old ones (Ma and Liang 2002).

The reservoir will be 5930 km in circumference at the completion of dam construction in 2009. It is predicted that about 9.8% of the bank will be unstable (Tao and Liu 2002). In July 2001 China proposed that geological disasters, especially the latent landslides caused by the reservoir filling with water, and water pollution in the reservoir area as well as other regions in the upper reaches of the Yangtze, must be prevented. The State Council ratified the geological disaster prevention overall plan in the reservoir area in February 2002, and China has implemented this plan since then (Tao and Liu 2002). The State Land and Resources Ministry is the authority in charge of a leadership group on this issue. Implementing this plan will protect the life and asset security of people residing on or affected by those landslides, and ensure ship navigation security and the quality of reconstruction of both urban and rural resettlement sites.

Prevention of water pollution

In November 2001, the State Council approved the 'Plan for Prevention and Control of Water Pollution in the Three Gorges Reservoir Area and the

Upper Reaches of the Yangtze', which was proposed by the SEPA. This plan aims to enhance the disposal capability of industrial and domestic sewage liquid and solid wastage. The SEPA has planned to invest about forty billion *yuan* over a duration of ten years. Starting in 2003, cities and towns will be strictly prohibited from discharging sewage that does not meet the required standard when the reservoir starts to be filled with water. Sewage and waste disposal facilities and equipment in the twenty cities and counties to be flooded will be improved.

Conclusion and implications

Environmental sustainability can be defined as the existence of the environmental conditions necessary to support human life at a specific level of well-being through future generations. The TGP has been regarded as a great and important ecological project. The Three Gorges reservoir can protect the vast areas and millions of people residing downstream of the dam from the threat of floods. Clean and renewable hydroelectricity generated by the project will make a contribution to the protection of the global environment, but the reservoir area will suffer from a range of irreversible and far-reaching environmental impacts. Permanent land inundation and a huge number of involuntary people to be displaced are the biggest unfavorable environmental consequences of the TGP.

Some secondary environmental impacts produced by human resettlement and urban relocation and reconstruction (including removal of industrial enterprises) have turned out to be crucial. Land lost, water and soil erosion, water pollution and geological hazards are closely related to rural and urban resettlement policy and schemes. The development of small- and medium-sized industrial and mining industries has further contaminated the water in the reservoir. The reconstruction of new cites and towns and the rebuilding of new roads and other basic infrastructure are likely to trigger severe geological hazards such as landslides and mud flows. Human economic activities, especially farming and reclamation of steep slopes and deforestation, have produced a greater impact on the increase of water and soil erosion, declining soil fertility, and sedimentation in the reservoir. Near resettlement projects, further exploitation of uphill land and traditional cultivation techniques, combined with deforestation for rural resettlement and firewood, have caused greater soil erosion and dramatic sediment accumulation downhill in the reservoir.

On the positive side, some constructive approaches and projects have been put into practice in order to minimize the number of people to be displaced and relieve the adverse impacts on the environment. On appropriate river banks, farmland protection engineering works are being constructed. Ecological farming systems have been established through the reform of traditional cultivation methods on slopes and the introduction of ecological farming techniques so as to reduce soil erosion and regenerate the agricultural production environment. A national-level reforestation project is being implemented in the reservoir area for the sake of environmental sustainability in the long run, but its implementation may negatively produce a throng of environmental emigrants, given the inherently fragile environment in the reservoir area and heavy pressure to displace and resettle a huge number of people associated with the TGP.

Future generations may face a declining land-to-people ratio and a deteriorating environment in the reservoir area. Thanks to the resettlement policy adjustments, resettling a large number of rural migrants beyond the reservoir area through distant resettlement schemes and shutting down some small industrial enterprises will undoubtedly relieve over-population and reduce pollutant sources in the reservoir area.

Some environmental impacts relating to rural resettlement and urban relocation have emerged, but others are uncertain and need time to surface. Some practical measures to control or reduce unfavorable changes to the environment in the reservoir area provide constructive implications for similar hydro projects in China and the world. The trend of environmental change in the reservoir area needs to be further observed and closely monitored.

References

Chen, G. J. (1987), 'The TGP area is already overpopulated and local resettlement will be a disaster', *Shuitu Baochi* (Soil and water conservation bulletin) (in Chinese), 7 (5): 42–46.

Chen, G. J., Xu, Q. and Du, R. H. (eds) (1995), *Studies on the Effects of the Three Gorges Project on Eco-environment and Countermeasures* (in Chinese), Beijing: Chinese Science Press.

Chi, W. J. (ed.) (1996), *An Outline of the Three Gorges Economic Development Area* (in Chinese), Beijing: China Three Gorges Press.

Changjiang Water Resource Commission (CWRC) (1996), 'Survey report on the physical flooding indices in Kaixian county' (in Chinese), prepared by the CWRC, Wuhan.

Changjiang Water Resource Commission (CWRC) (1997), *Study on Migration of the Three Gorges Project* (in Chinese), Wuhan: Hubei Science and Technology Press.

Changjiang Water Resource Commission (CWRC) and Changjiang Survey, Planning and Designing Institute (CSPDI) (1996), 'Kaixian rural resettlement planning report' (in Chinese), prepared by the CWRC and CSPDI, Wuhan.

Da, F. Q. (2001), 'Strategic measures for ecological rehabilitation in west China', *Qiushi* (Seeking the Truth Magazine) (in Chinese), 11: 53–55.

Du, R. H., Shi, D. M. and Yuan, J. M. (1994), *The Effects of Water and Soil Erosions on the Environment in Three Gorges Reservoir Area*' (in Chinese), Beijing: Chinese Science Press.

Du, Z. H. and Yan, G. A. (1999), 'Water and soil conservation and environmental improvements in the Three Gorges reservoir area', *Changjiang Liuyu ZiyuanYu Huanjing* (Resources and Environment in the Yangtze Basin), 8 (3): 299–303.

Fang, Z. Y. and Chen, Y. B. (2002), *Study and practice on the TGP environmental protection*, China Three Gorges Construction website: *www.cws.net.cn/journal/three_gorges/200209.*

Geographical and Resource Institute of the Chinese Academy of Science (GRICAS) (2000), 'Study on the resettlement and sustainable development in the Three Gorges reservoir area: rural resettlement and rural economic sustainable development' (in Chinese), prepared by the GRICAS, Beijing, May.

Gu, C. L. and Huang, C. X. (1999), 'Issues and countermeasures on the urban relocation and reconstruction in the Three Gorges reservoir area', *Changjiang Liuyu Ziyuan Yu Huanjing* (Resources and Environment in the Yangtze Basin) (in Chinese), 8 (4): 353–359.

Huan, W. W. (1996), 'Overview of the environmental protection in the dam site and reservoir area of the TGP', *China Three Gorges Project* (in Chinese), 4: 19.

Jiang, S. X. (1999), 'Assessment on environmental sustainability in the Three Gorges reservoir area', *Jingji Dili* (Economic Geography) (in Chinese), 19 (3): 64–69.

Kaixian Government (KG) and Changjiang Survey and Planning Institute (CSPI) (2000), 'Implementation planning of rural resettlement in the Changjiang Three Gorges reservoir area' (in Chinese), prepared by the Kaixian government and Changjiang Survey and Planning Institute, Wuhan, March.

Li, H. M. (2000), 'Population displacement and resettlement in the Three Gorges reservoir area of the Yangtze River central China', PhD thesis, The University of Leeds.

Li, J. and Luo, L. Z. (2001), *Construction of Eco-economy in West China* (in Chinese), Beijing: Nationality Publishing House.

Liang, F. Q. (1995), 'A preliminary approach on the issues of rural people's resettlement in the Three Gorges reservoir area', *Shanxia Gongcheng Jianshe* (China Three Gorges Construction) (in Chinese), 1: 47–49.

Liu, J. (ed.) (1999), Environmental construction plan in China (in Chinese), Beijing: Chinese Industry and Commerce Union Press.

Lu, X. X. and Higgitt, D. L. (2000), 'Estimating erosion rates on sloping agricultural land in the Yangtze Three Gorges, China, from caesium-137 measurements', *Cetena*, 39 (1): 33–51.

Lu, Y. M. (1996), The Three Gorges Project is a project which will improve the environment of the Changjiang River', *China Three Gorges Construction* (in Chinese), 4: 1–4.

Luo, M. and Xia, H. (2002), 'Impacts and countermeasures of TGP resettlement on the environment in the reservoir area', China Three Gorges Construction website: *www.cws.net.cn/journal/three_gorges/200209.*

Ma, Z. T. and Liang, H. H. (2002), 'Application of the sarma method to stability evaluation of jipazi slope, Yunyang County, Sichuan Province', *Dizhen Dizhi* (Earthquake and Geology) (in Chinese), 24 (3): 432.

Resettlement Expert Group of the TGP (REG) (1988), *Feasibility Report on the Resettlement Scheme of the TGP* (in Chinese), Beijing: Water Conservancy Press.

Research Institute for Protection of Yangtze Water Resources (eds) (1995), *Environmental Impact Statement for the Yangtze Three Gorges Project* (a brief edition), Beijing: Science Press.

Shi, D. and Yang, Y. (1989), 'Intensity of soil erosion in the Three Gorges area', in Study Group of Ecology and Environment for the Three Gorges Project (eds), *Atlas of Ecology and Environment for the Three Gorges Project* (in Chinese), 8–9, Beijing: Science Press.

Tao, J. L. and Liu, K. (2002), 'Important measures of protecting the environment—the 'two adjustments' and 'two preventions' of the TGP resettlement', China Three Gorges Construction website: *www.cws.net.cn/journal/three_gorges/200208.*

Wang, G. X. and Cheng, G. D. (2000), 'Eco-environmental changes and causative analysis in the source regions of the Yangtze and Yellow Rivers', *The Environmentalist*, 20 (3): 221–232.

Wang, L. M., Yang, Y. F. and Guan, Q. F. (2001), Environmental emigration stress of slope farmland in the three Gorges area, *Acta Geograpica Sinica*, 56 (6): 656–665.

Wang, R. S. (2000), 'The environment and resettlement of TGP', *Changjiang*

Liuyu Ziyuan Yu Huanjing (Resources and Environment in the Yangtze Basin) (in Chinese), 9 (1): 1–13.

Wang, Y. H. (1993), 'Analysis on income of migrants in China' (in Chinese), in X. Ma (ed.), Migration and Urbanisation in China, Beijing: New World Press, 36–45.

Wu, B. G. (1999), 'Speech of the deputy premier minister Wu Bang Guo on-the-spot meeting of the TGP distant resettlement', in Three Gores Project Construction Commission of the State Council of China (ed.), *Compiled Documents about the Scheme of the Distant-resettlement out of Chongqing Reservoir Section with TGP*, 14–18, Government document, Beijing.

Xu, Q. and Liu, Y. (1993), *Study on Environment Capacity for Migrants in the Three Gorges Reservoir Area* (in Chinese), Beijing: Science Press.

Yan, T., Yang, L. and Campbell, C. D. (2003), Microbial biomass and metabolic quotient of soils under different land use in the Three Gorges Reservoir area', *Geoderma*, 115 (1–2): 129–138.

Yang, C. Q. (2002), 'A decade of the dam: 10 years, 10 problems', Website: *www.probeinternational.org/pi/3g/index.cfm?DSP=content&Content ID=4043*

Yang, R. (1994), 'Sloping land resources in China and their utilisation models', *Natural Resources* (in Chinese), 1: 1–7.

Zhang, J. E. and Xu, Q. (1997), 'A few important issues regarding the sustainable development in the Three Gorges reservoir area that need to be resolved urgently', *Jingji Dili* (Economic Geography) (in Chinese), 17 (2): 54–59.

Zhang, X. (1996), 'Protection and utilization of water/land resources and improvement of environmental quality in the Three Gorges reservoir area', *Natural Resources* (in Chinese), 2: 6–13.

Zhu, N. (ed.) (1996), *Research on the Resettlement and Development of Three Gorges Reservoir Area* (in Chinese), Wuhan: Wuhan University Press.

Zhu, R. J. (1999), 'Transcript of Premier Zhu Rongji's speech at the State Council meeting on the Three Gorges project resettlement', in *Guidelines for the Distant Resettlement of Rural Resettlers in the Chongqing Reservoir Area of the TGP* (in Chinese), Compiled by Chongqing Resettlement Bureau, Chongqing: 1–16.

3

Ecologically Fragile Belts in Western China: Conservation and Rehabilitation

Wang Yiqian and Chen Yong

Introduction

Since 1999, the Western China Development policy has been implemented in order to achieve balanced development between the prosperous east and the vast underdeveloped western areas of China. In order to curb environmental degradation in western China, environmental conservation has been an essential part of this program.

Conservation and rehabilitation of the environment in ecologically fragile belts (EFBs) in western China is a major challenge. EFBs overlap with areas that are prone to natural hazards, have high concentrations of ethnic minorities and have high poverty rates. Conservation and rehabilitation of the environment in the EFBs is thus also significant in terms of poverty alleviation and the development of ethnic minority inhabited areas in China. To a great extent, socioeconomic development in western China has been constrained by the prevalence of EFBs.

EFBs in China

An overview of the eco-environment in western China

There are twelve provinces, municipalities and autonomous regions in western China, covering an area of 6.85 million km^2, accounting for 71.4% of the total landmass in China (Figure 3.1). The population in western China reached 367 million in 2002, making up 28.8% of the whole country. Gross domestic product (GDP) amounted to 1988.6 billion *yuan* in 2002, seventeen percent of the national total.[1] Situated on the highest and second highest terrains of the overall landforms of China where a number of major rivers originate, the uneven relief of western China renders it prone to soil erosion (Figure 3.2). Hence the ecological stability of western China affects the ecological safety of central and eastern China.

Figure 3.1 Provinces in western China

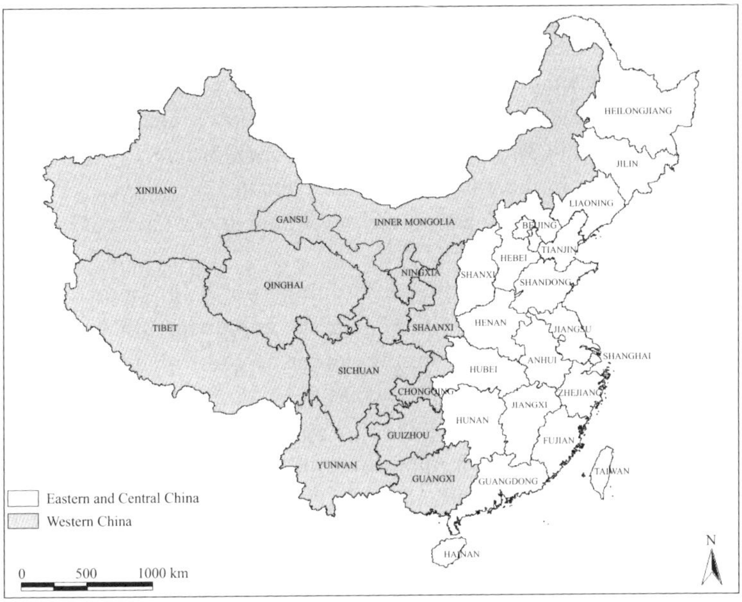

Figure 3.2 Elevation map of China

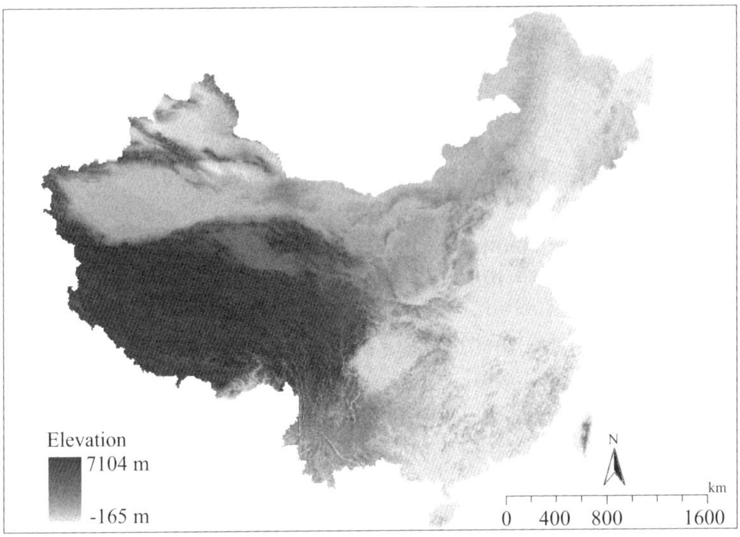

Definition of an ecologically fragile environment (EFE)

In recent years, Chinese environmental scholars have paid great attention to ecologically fragile environments (EFEs), especially in the Western China Development Program. An EFE is an environment with high sensitivity and instability (Yang et al. 1992; Liu 1993; Zhao 1999; Zhao and Liu 1996; Huang 2003; Xiao 2003). Sensitivity refers to cases where any disturbance of a major element in the environment may result in a series of reactions that result in a negative impact on the environment. Instability has strong spatial and temporal dimensions.

Negative impacts on an EFE may include pollution and damage to resources (Zhao 1999) as well as ecological imbalance. Such an EFE is a continuous area where degradation cannot be controlled under present economic and technological conditions, as the ecosystem is unable to resist disturbance (Huang et al. 2003).

The eco-environment is an integration of matter, energy, living space and the bases of socioeconomic activities which support people's life and development. Since an EFE includes pollution and damage to resources, its characteristics go beyond high sensitivity and instability to include the following five aspects:

1. A low ability to maintain the stability of the external environment for human survival and development;
2. A low ability to buffer waste released by human beings in their economic activities;
3. A weak capability to degrade toxic waste;
4. A weak resistance to disturbances of ecosystems;
5. A low capability to rehabilitate damaged ecosystems.

An ecologically fragile belt (EFB) is closely related to the biophysical features, social conditions and economic development of a region. It may be a result of both environmental evolution and human socioeconomic activities. Its fragility is not only indicated in the instability of its internal structure and sensitivity to external disturbance, but also by its low ability to support human socioeconomic activities. Assessment of an EFB may include evaluation of its present situation, development trend and stability.

Classification of EFBs in China and their location

Human impacts on the environment differ widely in China. EFBs in China may be divided into seven areas according to topographical conditions and climatic factors, as shown in Figure 3.3 and Table 3.1 (Zhao 1999). They may be further divided into a number of sub-belts based on landscape characteristics, soil features and other factors.

Figure 3.3 Distribution of ecologically fragile belts in China

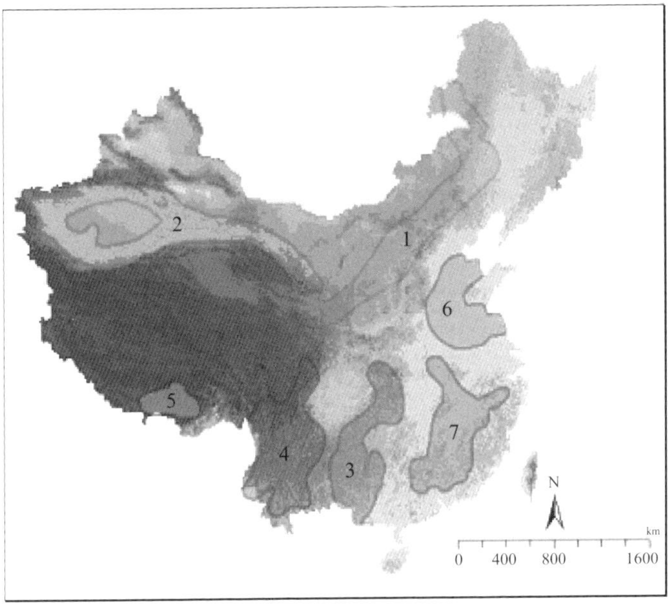

Source: Adapted from Zhao (1999).

Table 3.1 and Figure 3.3 show that five of the seven major EFBs in China are located in western China, i.e., away from the developed coastal area. The two exceptions (#6 and #7) are in northern China and the hilly areas of southern China. The fragile environment in western China may not only be a major constraint for economic development in these areas, but may also threaten the nation's existence in the future (Shi and Liang 2002). Over-use of water for irrigation, over-grazing, irrational mining development and over-harvesting of trees by local inhabitants may result in further degradation of the inherently fragile environment (Ge 1990).

Major problems facing EFBs in Western China

There are a number of environmental problems facing the ecologically fragile regions in western China such as soil erosion, desertification, land degradation, lack of water resources, receding of glaciers, reduction in biodiversity, degradation of grassland, aggravation of natural disasters, urban and industrial pollution and over-mining (Zhu 1991; Chang et al. 1999).

Table 3.1 Classification of ecologically fragile belts (EFBs) in China

Code and name of EFBs	Main distribution areas	Principal factors	Indicators
1 Semi-arid and semi-humid areas in northern China	Striped-shaped. From the western border of the Daxing'anling Mountains in the north to the western part of Keerxin sandy land in the east, and the eastern part of Hexi Corridor in the west; width from Maowushu sandy land to Loess plateau can be more than 300 km.	• Little rainfall • Irregular precipitation • Impacts of evaporation and precipitation on land use	• 400 mm annual rainfall • 400 mm precipitation assurance rate:>50% • 350 mm precipitation assurance rate: >50% • Aridity: 1.5–2.0
2 Semi-arid areas in northwestern China	Circular and striped-shaped. Includes the circular region of the southern slope of Tianshan mountains and the northern slope of Kunlun mountains and the long strip stretching from Hexi Corridor to Lake Luobubo.	• Shortage of water • Unstable supply of water resource • Wind erosion • Sand accumulation • Over-cultivation	• Water runoff losing area • Runoff variation by +50% or –50% • Vegetation cover in surrounding areas <10% • Area of windbreak network <10%
3 Lime rock mountains in southwestern China	Extending southwards from the eastern part of Daba Mountains to the convergence areas of Sichuan, Chongqing, Yunnan, Guizhou provinces and Guangxi Autonomous Region.	• Dissolution • Hydraulic erosion	• Limestone mountains • Vegetation cover <30%
4 Mountainous areas in southwestern China	Centralised in the central section of the Hengduan Mountains, mainly arid valleys and basins.	• Hydraulic erosion • Drought • Over cultivation • Over harvest of timber • Over-grazing	• Moderate and greater cutting by rivers • Hydraulic erosion • Aridity >1.5 • Vegetation cover <30%
5 Qinghai-Tibetan Plateau	Distributed over the middle region of the Yanung Zangbo River.	• Little and irregular precipitation • Impacts of evaporation and precipitation on land use • Over cultivation and over-grazing	• Annual rainfall: 400 mm • 400 mm precipitation assurance rate >50% • 350 mm precipitation assurance rate >50% • Aridity 1.5–2.0
6 The plain areas in northern China	Discontinuous distribution approximately from Huayuankou in Henan Province to the alluvial plains of the Yellow River, extending to seaside plains.	• Poor drainage • Windstorms • Wind erosion	• Underground water depth: more than 3 m deep • Mineralization degree of underground water >2g/l • Old and new sandy land in the channels of the Yellow River • Vegetation cover<30%
7 Hilly areas in southern China	Equivalent in distribution to the red basins and hilly areas in the southern bank of the Yangtze River; discontinuous distribution.	• Over cultivation • Over harvest of firewood • Hydraulic erosion	• Vegetation cover <30% • Red soil hills and mountains • Storms

Source: Adapted from Zhao (1999: 16–42).

This study focuses on a few environmental problems that are affecting the ecology of larger regions.

Serious soil erosion

Soil erosion may be the most serious environmental problem in China. It is reported that the annual total volume of eroded soil amounts to five billion tons, of which two thirds is from western China[2]. A remote sensing survey from the Ministry of Water Conservancy of China shows that about 3.56 million km^2 of land in China is suffering from soil erosion. Of this, 1.65 million km^2 results from water erosion and 1.91 million km^2 from wind erosion. Soil erosion in western China includes wind erosion in northwestern China, water erosion in southwestern China and freeze-thaw erosion on Qinghai-Tibetan Plateau (Photo 3.1).

Soil erosion in EFBs in western China has the following characteristics:

1. Soil erosion occurs mainly in Yunnan, Guizhou, Sichuan, Chongqing and Hubei provinces in the upper Yangtze River and Shanxi, Shaanxi, Inner Mongolia, Gansu and Ningxia provinces in the middle Yellow River.
2. Soil erosion becomes more serious as one goes west. Land suffering from water erosion in eastern, central and western China is respectively 90,000 km^2, 490,000 km^2 and 1.07 million km^2.
3. Northwestern China suffers most seriously from wind erosion, as in Xinjiang, Inner Mongolia, Gansu and Qinghai provinces.
4. The areas adjoining the Great Wall and the agri-pastoral areas in Xinjiang suffer from both water erosion and wind erosion.

In contrast with areas that suffer from soil erosion in China as a whole, an increasing area of land in western China is exposed to water and wind erosion due to reduction in vegetation cover, less rainfall and more droughts in arid areas in recent years. Over-harvesting of trees, over-cultivation of arable land, irrational exploitation of inland water resources and neglect of soil conservation practices during implementation of civil engineering may be causes for soil erosion. About 450,000 km^2 of land on Loess Plateau, which comprises seventy percent of the total area of Loess Plateau, is prone to soil erosion. As the origins of sediment in the Yellow River, Loess Plateau has suffered the most serious soil erosion in the world. In the upper and middle Yangtze River, about 550,000 km^2 of land (thirty-five percent of its total area) is afflicted with soil erosion. Soil erosion may not only present a threat to the life support system, but can also lead to siltation in rivers and floods downstream. Situated at an elevation of over 3000m, Qinghai-Tibetan Plateau has about 1.04 million km^2 of land threatened with freeze-thaw erosion.

Photo 3.1. Sediment accumulation due to serious soil erosion in Xichang, Sichuan province

Deterioration of desertification

Desertification refers to the degradation of land in arid, semi-arid or dry semi-humid areas through water erosion, sediment accumulation and salination. As a big country that suffers from serious desertification, China has 2.622 million km^2 of land, 27.3% of its total area, which is threatened by desertification. About 2460 km^2 of land is desertified each year with an economic loss of fifty-four billion *yuan*. The provinces that have the largest area of desert land are Xinjiang (1.044 million km^2), Inner Mongolia (659,000 km^2) and Tibet (436,000 km^2) (Wang and Zhang 2002). Desertification has been an important environmental factor affecting economic development in China. At present China is still in a state of overall deterioration despite regional improvement with regard to desertification. The resulting natural disasters are becoming more frequent and serious, and the losses greater.

Tibet has an area of 307.11 million ha of desertified land, making up 17.03% of its total landmass (Photo 3.2). In addition, there is another 20.17 million ha of land that is threatened by desertification (Development and Planning Commission of the Tibet Autonomous Region, 2003). As we noticed during our investigation in Tibet, the Tibetan Plateau has the basic conditions for the occurrence of desertification. This condition may be exacerbated by global warming. An increasing gap between the low carrying capacity of the land and a large population of people and livestock might lead to over-exploitation of land resources, which will further aggravate desertification.

Photo 3.2. The wandering dunes on grassland in Tibet

There is an increasing area of land in the karst region in Yunnan, Guizhou and Guangxi provinces that is suffering from rock desertification where slopes are stripped of soil or vegetation. As of 2003, about 70,000 km^2 of land is afflicted with rock desertification. Rock desertification, general desertification and soil erosion have been the three main ecological hazards in China in recent times (CCTV News 2004). Guangxi province has 4.72 million ha of rocky mountains, of which three million ha has been rock desertified, accounting for one third of its total land area. Despite much effort to control rock desertification, the area that suffers from it is increasing at a rate of three to six percent every year. In Guizhou province, about 1800 km^2 of land is being rock desertified each year and about seventy-six km^2 of arable land is lost each year. The ratio of area of rock desertified land in Yunnan province has increased from seven percent in the early 1950s to thirty percent at present. With the implementation of the 'Grain to Green' program and natural protection program, environmental degradation in karst areas has been mitigated to some extent but the overall situation of rock desertification remains serious.

Scale of forest destruction

Forests have played an essential role in environmental conservation. Any action to damage forests may contribute to soil erosion and the occurrence of floods and dust storms. Except for Shaanxi, all provinces in northwestern China have a forest coverage rate below the average level in China. The forest coverage rate is 0.35%, 0.79%, 1.54%, 4.33% and 5.84% in Qinghai,

Table 3.2 The economic loss caused by land desertification in Tibet

Location	Damage by desertification					Annual economic loss (10^4 *yuan*)	Density of economic loss (*yuan*/km^2)	Average per capita economic loss (*yuan*/person)
	Arable land (10^4ha)	Grassland (10^4ha)	Villages (No.)	High way (km)	Water channel (km)			
Lhasa	0.33	5.33	2	21.0	3.0	981.0	399.4	26.3
Linzhi	0.01	7.02	8	11.3	13.0	580.2	51.7	42.7
Changdu	0.06	0.42	5	0.0	1.2	510.4	47.9	9.8
Shannan	0.37	1.33	2	20.0	5.0	984.0	126.5	33.2
Rikaze	2.17	317.60	450	240.0	165.0	48 000.0	2385.7	823.5
Naqu	0.11	1026.87	181	200.0	0.0	30 248.1	7874.3	953.6
Ali	0.12	340.27	100	30.0	6.0	5000.0	171.9	764.5
Total	3.07	1698.84	784	522.3	193.2	86303.7	718.0	377.1

Source: Dong (2001: 145–151).

Xinjiang, Ningxia, Gansu and Tibet respectively. The total area of forests in northwestern China (excluding Shaanxi province) is only four million ha, accounting for only three percent of the total area of forests in China, where forest coverage rate is only 1.75%. Irrational exploitation of water resources and over-harvesting of trees may be some of the major reasons underlying this situation.

Southwestern China enjoys abundant forestland, but forests have been substantially destroyed (Photos 3.3 and 3.4). Sichuan province, for instance, once experienced a decline of forest coverage rate by more than sixty percent from the 1930s to the 1980s. In the 1950s forest coverage rate in Sichuan was only twenty percent, but it declined to its lowest level of nine percent in the 1960s. Data show that the devastating flood on the Yangtze River in 1998 was closely related to loss of forests in upstream regions (Yang and Chen 2003).

Thanks to the 'Grain for Green' program and the natural forest protection program, forest coverage rate in EFBs has increased. However, there is a long way to go to achieve a remarkable ecological effect in EFBs.

Water scarcity

Outside of the Sichuan Basin, most areas in western China face water scarcity. In northwestern China where annual rainfall is often below 400 mm and average annual evaporation for years has been as high as 1200 mm, there is a big gap between supply and the demand for water resources. Lakes are drying up and some rivers have ceased to flow due to irrational exploitation of water resources. The scarcity rate in recent years in northwestern China is 8.7%. It is even higher in some places like Guanzhong district in Shaanxi, oasis district in Xinjiang, Hexi corridor district in Gansu and the Shiyang

Photo 3.3. Cleared forest in western Sichuan

Photo 3.4. Soil erosion after deforestation in western Sichuan

River watershed in Xinjiang (Table 3.1) (Yang et al. 1994). Although rainfall in southwestern China is as high as 1000mm, there is a low capacity for water storage since mountains and hills are sporadic. It is too difficult to build reservoirs in mountainous terrain.

Qinghai-Tibetan Plateau enjoys a great number of lakes, covering a total area of 36,900 km^2, making up fifty-two percent of the total area of all lakes in China. Many lakes are facing the problem of becoming dry (Photo 3.5). In addition to climatic change and decline of ground water levels, one key factor in water scarcity is increased human activities.

Photo 3.5. Dry salt lake in Tibet

Degradation of grassland

At present over ninty percent of grassland in China is suffering from degradation. The current overgrazing rate is on average thirty-six percent, which is eighteen percent higher than that in the 1980s. Degradation of grassland in northwestern China is more serious than in any other place in China. For instance, the degradation rate in Ningxia, Shaanxi, Gansu and Tibet is, respectively, as high as 97.37%, 58.55%, 45.17% and 30.36%. The main cause for this is overgrazing. Degraded grassland is expanding at an annual rate of 200 ha and the loss of grassland each year amounts to 650,000–700,000 ha[3]. One of the five major grasslands in China, in northwestern Sichuan, has experienced a rapid degradation of grassland ecosystems since the 1960s. At present desertified land in this region is expanding at a drastic rate (Photo 3.6). According to a survey by Sichuan Agricultural Academy, about four million ha of grassland has suffered from a rodent epidemic for over a decade, which may lead to an annual loss of forage grass reaching one billion kg. Moreover, over-harvesting of herbs will have a negative impact on grasslands (Photo 3.7).

Approaches to conservation and rehabilitation of ecologically fragile environments

Ecological recovery of EFBs is essentially a process of natural progression. This process is usually slow and there may be some irreversible damage. Conservation and ecological reconstruction of EFBs in western China are neither simply ecological rehabilitation, nor a technical problem. They are

Photo 3.6. Degraded grassland in Ganzi, Sichuan

Photo 3.7. Rat habitats on grassland in Tibet

processes related to the natural-technical-social complex. All factors concerning the physical environment and human society should be considered.

Much work has been carried out on studies of conservation and rehabilitation of EFBs (Yan et al. 2000; Shi and Liang 2002). There are many new ideas and insights on this issue, including: reinforcing environmental management, establishing compensation mechanisms in watershed, promoting the development of eco-agriculture, establishing more nature reserves, controlling soil erosion and preventing desertification. However, there may be different underlying causes of environmental degradation for different EFBs and it would be very hard to find a common solution to solve all problems. According to our experience in western China, much effort should be taken to adjust industrial structure and energy use structure, redistribute populations and alleviate poverty in addition to solving the technical problems of ecological rehabilitation. It is important to make full use of the self-repair ability of the ecosystem. Considering the present situation of EFBs in western China, we may focus on the following aspects in the near future.

Promoting eco-tourism

In the areas involved in the 'Grain to Green' program, farmers' employment is a big problem. All surplus labor needs to be transferred from agriculture to non-agricultural activities. The policy of compensation for farmers who have lost their arable land has given us an opportunity to seek alternative industries. Any alternative industry or activity should be in line with the aim of ecological reconstruction and there should be no ensuing damage to the ecosystem.

This study suggests eco-tourism as a pioneer industry for the ecological economy. The rise of eco-tourism may promote regional development, including development of trade in transportation, commerce and recreation. Eco-agriculture and bio-medicine industries may also need rapid development.

EFBs in western China enjoy abundant and various eco-tourism resources. There are snow-covered mountains, ice peaks, deep valleys and rare stones as well as big rivers, waterfalls and hot and cold springs. Forests, grassland, rare animals and plants and magnificent flowers are also attractive to tourists. At present, three different eco-tourism zones have been formed. In 2002, western China received over 200 million tourists generating an income of over 100 billion *yuan*.

Development of eco-energy

Since the implementation of the program of natural forest protection, EFBs in western China have begun to face the problem of fuel scarcity. The main energy source for domestic use in EFBs has been wood, turf and straw. About 140 million people in thirty-five million households in the areas covered by key programs of ecological reconstruction have been dependent on wood, turf and straw for cooking food. These farmers would burn 420–560 million m^3 of bushes removed from 233,000 km^2. In pastoral areas, cattle manure has been the main fuel for cooking food. Clearing cattle manure from grasslands may deplete organic matter and lead to grassland degradation. Biogas can be used in rural areas in eastern China, but it is very hard to make it viable in the highlands of western China, where the dry and cold weather affect the fermentation of organic matter.

In western China there is a great potential for developing eco-energies, such as hydropower, wind energy, solar energy and thermal energy. For example, Ganzi Prefecture in the western part of Sichuan is abundant in hydropower resources as a number of big rivers flow through it. Its potential storage of hydropower is 37.29 million kw, making up a quarter of the total possible storage of hydropower in Sichuan. Sichuan provincial government has planned to promote the development of small- and medium-scale hydropower stations. Once the maintenance work on the electric network in rural areas is completed, energy problems in rural areas could be solved. As of 2003, about 1.5 million households have started to use electricity instead of wood fuel for cooking. Thanks to the use of electricity, over four million m^3 of wood could be saved each year, translating into the protection of over 40,000 ha of forestland per annum.

Promotion of ecological migration

A typical feature of EFBs is low carrying capacity in terms of human population. Ecological migration may provide a lever to adjust population pressure. In the EFBs in western China, many places are not suitable for human inhabitation. Living in harsh environments, many people will never be able to rise above the poverty line. Moreover, over-exploitation of local resources may aggravate environmental degradation. Resettlement of these people from fragile environments to places with better environmental conditions will relieve the high ecological pressure on these environments, so that the already damaged ecosystem may recover.

There are three types of places where resettlement programs can be practiced:

1. Places where the primary environment is not suitable for people to live in, such as at high altitudes with low temperatures and water scarcity, or severely arid areas with intense water scarcity;
2. Places where the ecosystem is in the process of regressive development, such as grasslands that have suffered serious desertification, belts that are vulnerable to mountain hazards like landslides and debris flows and areas that suffer from severe soil erosion;
3. Nature reserves where all inhabitants, particularly those in the core areas, may be resettled outside so that landscapes and biodiversities could be better conserved.

As a large province in western China, Sichuan has an estimated one million people who need to be resettled to achieve poverty alleviation. It is planned that about 600,000 people in 150,000 households will be resettled before 2005

(Sichuan Provincial Office for Western China Development 2001). These are mostly ethnic minorities from Ganzi, Aba and Liangshan prefectures in western Sichuan. In the implementation of resettlement programs, several issues should be of much concern, including: the willingness of migrants to uproot, assurance of the interest of migrants, provision of migration funds, provision of arable land and community integration and religious conflict (Wang and Wang 2003).

More effort is needed for anti-poverty campaigns

The aim of the Western China Development Program is to narrow the regional gap in development. Up till 2000, there were still thirty million people living in poverty in China. Around half of these people are from western China. Of all the 592 poverty-stricken counties identified by the State, 366 counties are situated in western China. When linking distribution of poverty-stricken counties with EFBs, it is not difficult to find that the geographical distribution of impoverished peoples is highly correlated to EFBs in China. Of all counties in the EFBs, about seventy-six percent are poverty-stricken, accounting for seventy-three percent of all poverty-stricken counties in China. As for the population living in EFBs, seventy-four percent are living in poverty-stricken counties, comprising seventy-eight percent of the total population in all poverty-stricken counties (Figure 3.4). Most poverty-stricken counties in western China are either in remote mountainous counties or ethnic minority counties. Of all 366 state poverty-stricken counties, 258 are mountainous counties and 210 are ethnic minority counties.

In the poverty-stricken areas of western China, people are used to ways of production that pursued increase in quantity instead of quality. This way of production may lead to over-exploitation of natural resources when there is a rapid population growth. As a result, environmental degradation and decline of land productivity is inevitable. Once there is greater pressure on the land from a larger population, more serious degradation would occur, which may in turn lead to more poverty. Thus, people would be trapped in a vicious circle of 'destruction of ecosystem–poverty–more destruction of ecosystem–more poverty' (Li et al. 1997). Since most poverty-stricken counties in western China are situated in EFBs, anti-poverty campaigns on the basis of environmental conservation and reconstruction may tackle the two problems at once.

Policy approaches

With regard to environmental degradation in western China and its possible impact on the sustainable development of the whole of China, the central government of China has proposed an urgent task to strengthen environmental

Figure 3.4 Distribution of poverty-stricken counties identified by the state

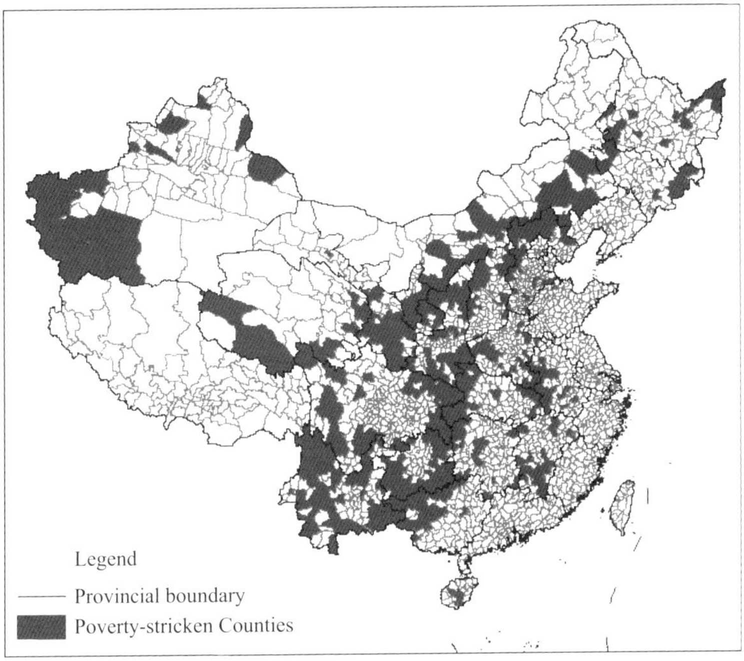

Source: State Council Leading Group Office of Poverty Alleviation and Development (2003: 56–58).

conservation and rehabilitation. In order to achieve this task, a series of programs have been implemented, such as the 'Grain for Green' program, and programs of natural forest protection, desertification control, degraded grassland restoration, soil conservation and reconstruction of nature reserves. Of all these programs, the 'Grain for Green' program is arguably the most effective and has been a major part of the plan for environmental reconstruction. As of 2003, the acreage of converted cultivated land has reached 15.13 million ha, of which 7.2 million ha have been converted to forestland and 7.93 million ha are reforested wasteland (*Xinhua News Agency* 1 December 2003). The overall environmental condition in China is, however, still serious. More areas are suffering from ecological degradation and more places are facing serious environmental problems than ever before. There are cases where environmental destruction is occurring at the same time as ecological restoration, though the degradation is happening at a faster rate than the restoration. In some places, more natural disasters are occurring due to the damaged ecosystem (*Xinhua News Agency* 22 October 2003). The function of the natural ecosystem is decreasing and the resilience of the environment against natural disasters is poor (State Environmental Protection Bureau 2003).

Conclusion

Fragile environments may have been influenced by physical factors, human activities and their interactions in the past. The fragility of the environment is characterized by instability of its endogenous structure and sensitivity to external disturbance. Moreover, fragile environments have a low resistance to environmental pollution and resource depletion. Equally important, the vulnerability of the environment is characterized by a low capability for supporting human socioeconomic activities. This capability at least includes the stability, resilience and rehabilitation of ecosystems.

The Chinese government has invested greatly in ecological reconstruction in recent years, yet the area suffering environmental deterioration is expanding and becoming worse. The integral function of ecosystems is decreasing and its capacity to prevent natural disasters weakening. Regions with critical fragile eco-environment in China are mainly found in western China. These areas need to be strictly controlled and protected.

The main problems of EFBs in western China include: serious soil erosion, worsening desertification, deforestation, water scarcity and grassland degradation. Protection and rehabilitation of the fragile areas in western China is not merely a technical issue and cannot be simply regenerated; it is an integral process of nature, techniques and society. In regard to ecological or environmental reconstruction in the western regions of China, it is suggested that: it is imperative to stop unsustainable production patterns by which people completely degrade the natural resources and environment; population growth should be compatible with the human carrying capacity of the environment; and the poverty-stricken people in the regions with fragile environments need to be assisted in improving their living and development capacity. The development of eco-tourism and the eco-energy industry and the implementation of environmental migration and poverty assistance programs may be important approaches to protecting and reconstructing the environment in west China.

Notes

1. US$ 1 = RMB 8.27 *yuan* in December 2003.
2. See *www.peopledaily.ac.cn/GB/jingji/1037/2167467.html.*
3. See *http://cydgn.cyol.com/gb/cydgn/2003-08/18/content_717069.html.*

References

CCTV News (2004), 'Daily News Broadcast: Rock desertification is increasing at an alarming rate in South-western regions of the country', January 1.

Chang, X. L., A. F. Zhao, and S. G. Li, (1999), 'Spatial-temporal scale and hierarchy of vulnerable ecotone', *Journal of Desert Research* (in Chinese), 19 (2): 115–119.

Development and Planning Commission of the Tibet Autonomous Region (2003), 'An overview of eco-environmental protection and reconstruction in Tibet' (in Chinese), *Tibet Daily*, September 15.

Dong, Y. X. (2001), 'Study on prevention of desertification and engineering problems in the Tibet Autonomous Region', *Natural Resource Journal*, 16 (2): 145–151.

Ge, Q. S. (1990), 'Studies on characteristics of ecologically fragile belts in China', *New Ideas in Geography* (in Chinese), 5 (2): 11–17.

Huang, C. M. and N. S. Ai (2003), 'Characterization of fragile eco-regions in southwest China', *Resources and Environment in the Yangtze Basin* (in Chinese), 12 (5): 467–472.

Li, Z., R. M. Chen and L. Gao, (1997) *Studies on the Relationship Between Development and Environmental Change* (in Chinese), Taiyuan: Shanxi Economic Publishing House.

Liu, Y. H. (1993), *A Preliminary Study on Ecologically Fragile Environments: Studies on Comprehensive Control of Eco-environmental Degradation and Restoration Techniques* (in Chinese), Beijing: Beijing Science and Technology Press.

Shi, D. M. and Y. Liang (2002), 'Assessment and protection of fragile eco-environments in China', *Journal of Water and Soil Conservation* (in Chinese), 16 (1): 6–10.

Sichuan Provincial Office for Western China Development (2001), A Report of Studies on Ecological Migration in Sichuan (in Chinese), unpublished paper.

State Council Leading Group Office of Poverty Alleviation and Development (2003), *An Overview of the Development-oriented Poverty Reduction Program for Rural China*, Beijing: China Financial and Economic Publishing House.

State Environmental Protection Bureau (2003), 'The environmental problems in China are still severe', *http://news.xinhuanet.com/zhengfu/2003-10/23/content_1137709.htm*.

Wang, F. and Y. Q. Wang (2003), 'Ecological migration and sustainable development in upper Yangtze River', *Population and Economics* (in Chinese), 2: 63–68.

Wang, H. Z. and J. Z. Zhang (2002), 'Advance on vulnerable ecological region and its restoration technique', *Rain Fed Crops* (in Chinese), 22 (4): 233–236.

Xinhua News Agency (2003), 'A total area of 227 million land has been converted from arable land to forestland', December 1.

Xinhua News Agency (2003), 'China is still facing a serious environmental situation', October 22.

Xiao, D. N. (2003), 'Ecological reconstruction and landscape planning of ecological fragile zones', *Journal of Desert Research* (in Chinese), 23 (1): 6–11.

Yan, S. M., Y. C. Zhu, and X. H. Han (2000), 'The critical areas obstructing sustainable development in ecologically fragile regions and solutions', *Ecological Economy* (in Chinese), 11: 18–24.

Yang, D. G. and G. J. Chen (2003), *Ecological Restoration and Sustainable Development in Upper Yangtze River* (in Chinese), Chengdu: Sichuan University Press.

Yang, Q. Y., X. H. Liu and G. D. Li (1994), 'Critical environmental situation in middle Yellow River', *Yunnan Geographic Environment Research* (in Chinese), 6 (1): 32–43.

Yang, Q. Y., Y. L. Zhang, and G. D. Li (1992), 'Situation of fragile environments and areas of severe fragility in China', *Geographical Research* (in Chinese), 11 (4): 1–9.

Zhao, Y. L. (1999), *Distribution Pattern of Ecologically Fragile Environments in China and Comprehensive Restoration* (in Chinese), Beijing: China Environmental Science Press.

Zhao, Y. L. and Y. H. Liu (1996), 'The decision of the region of fragile environment in China and the study on the relationship between fragile environment and poverty', *Advance in Earth Sciences* (in Chinese), 11 (3): 245–251.

Zhu, Z. D. (1991), 'Ecologically fragile belts and desertification in China', *Journal of Desert Research* (in Chinese), 11 (4): 11–22.

4

Ecological Migration in the Karst and High, Cold Mountainous Areas of South China

Cang Ming

Ecological migration or, alternatively, population movement for better ecological conditions, usually refers to migration motivated by a desire to alleviate ecological pressure. Before the 1980s, Chinese society was in widespread poverty.[1] According to the 1978 figures published by China's National Bureau of Statistics, China's population in absolute poverty at that time amounted to 250 million.[2] In 1979, the Chinese leader Deng Xiaoping launched an economic restructuring program in China's rural areas, providing long-term leases of land to each household. By 1985, the poverty-stricken population had been reduced by half, reaching 125 million. Since 1986, China's central government has made a special commitment to anti-poverty efforts through investing in and supporting the economically underdeveloped areas in a planned and organized way. In China this is called *fupin*, meaning poverty alleviation. The impoverished rural population fell from 125 million in 1985 to eighty million in 1993. The poor population is concentrated in China's peripheral or marginal areas, mostly in southwest China's karst and high, cold mountainous areas and northwest China's loess plateau. Several million people in these two areas have lost their very conditions of survival, living in absolute poverty and falling victim to ecological disasters. To address this problem, the Chinese government decided to launch a migration program to move these people from their original places of extremely adverse natural conditions to locations where the natural conditions were much better. This process allows migrants to climb out of poverty and at the same time to relieve the ecological pressure exerted by these populations. The Chinese authorities and media usually call this kind of population movement ecological migration, migration for poverty alleviation, or development-oriented migration. Large-scale ecological migration has been carried out against such a background in the karst and high, cold mountainous areas in south China's Guangxi, Guizhou, Guangdong and Yunnan.

Ecological background of migration

Ecological changes in the Karst mountainous areas

China has a total karst land area of 1.368 million km^2, ranking second in the world. It is distributed in the seven provinces or autonomous regions of Guangxi, Guizhou, Yunnan, Guangdong, Sichuan, Hunan and Hubei, of which the concentrated and connected areas are found in Guangxi, Guizhou, Guangdong and Yunnan. Guizhou and Yunnan have the largest karst areas, which reach 79.27% and 37.81% respectively (Yang 1998: 6). The subtropical karst areas that receive rich rainfall and have good vegetation cover are endowed with beautiful landscapes and are becoming tourism getaways. For example, Guangxi's Lijiang River and Guizhou's Huangguoshu Waterfall are well-known destinations for tourists. However, in the karst areas there is undulating terrain, high mountains and deep valleys, precipitous slopes and little top-soil, scarce flat land with thin and poor soil coverage and fragile vegetation that, once destroyed, is very hard to recover.

Guangxi

In the Qing dynasty, Guangxi's large rocky mountainous areas were covered with dense forests, maintaining a better ecological environment than the present situation. According to *Xilin Xianzhi* (Annals of Xilin County), compiled in the Qing dynasty in the reign of Kangxi Emperor, Xilin County at that time was 'covered with dense and deep forests in the mountains'. According to *Guangxi Tongzhi* (General Records of Guangxi) compiled in the Qing dynasty in the reign of Jiaqing Emperor, Nandan and Donglan counties were endowed with many deep valleys and lush bushes and trees. In Debao (which was De'an Prefecture in the Qing dynasty) there were 'deep mountains and dense forests', and it was 'thickly wooded with tall trees', and 'the towering trees look luxuriant'. *Guishun Zhouzhi* (Annals of Guishun Prefecture) reads that the local 'bushes and trees are emerald green'. Guishun was present-day Jingxi County. *Longsheng Tingzhi* (Annals of Longsheng Prefecture) compiled in the Qing dynasty in the reign of Daoguang Emperor states that 'Longsheng is embraced by the mountains with bamboo groves and deep valleys'.

The 1989–1993 figures of the fourth national forests resources inventory show that Guangxi's forest coverage rate was only 25.34% (Editorial Board of *Forests in China*, 1995: 231), and the percentage was even lower in karst areas: there were twenty-seven rocky counties and cities where the forest coverage rate was below thirteen percent, of which that of eight counties was below ten percent, where barren rocks are exposed, suggesting serious ecological destruction. The massive deforestation has led to serious soil erosion. The 1993 statistics show that Hechi and Bose of karst rocky areas had the largest area of soil erosion, which were respectively 7332 km^2 in Hechi and 6329 km^2 in Bose (Xiao 1996: 228). Soil erosion gravely threatens the ecological environment for agriculture. The cultivated land of those two areas accounts for only ten percent of the local land area. Therefore the people vividly call it 'ninety percent of rocks versus ten percent of soil'. The largest rocky mountainous areas are found in Du'an Yao Autonomous County, where 260,000 people have no more than 0.5 *mu* (one *mu* = 0.0667 ha) of cultivated land area per capita. Most of the cultivated land, sometimes as small as a rice bowl or a wooden dipper, was found between the rocks on the mountainsides, with very few connected tracts of land. The soil is impoverished and forms a very thin layer of usually only twenty cm deep. Apart from this, there is little rainfall in spring and winter across these rocky areas, resulting in serious water shortages. For example, in 1993, 180,000 people in Du'an Yao Autonomous County, 100,000 people in Mashan County, and 100,000 people in Dahua Yao Autonomous County had difficulties in obtaining drinking water (Xiao 1996: 177). The living conditions are so bad there that whatever support is rendered by the government will likely fail to help the local people cast off the shackles of poverty.

Guizhou

In the Ming dynasty, the state of Guangshun in the Guizhou karst region had many trees, Xu Xiake Travels said: 'The big trees are dense, and the streams flow south' (Zhu 1985). Even in the Qing dynasty, there were dense forests in Guizhou. According to Guizhou Tongzhi (General Records of Guizhou) compiled in the reign of Qianlong Emperor, in northeast Guizhou the following conditions existed: 'in the counties under the jurisdiction of Sinan, Zhenyuan and Tongren prefectures, with a circumference of more than 600 li (one li = 1/2 km), there were dense forests with deep, dark old trees. Most of the trees are pine, tung, oil-tea camellia, white wax trees and Chinese fir as well as cypress, maple, white poplar, nanmu and Chinese catalpa. According to Xingyi Fuzhi (Annals of Xingyi Prefecture) compiled in the reign of Guangxu Emperor, in Guizhou's southwest region, i.e. in Xingyi, Xingren, Pu'an and Zhenfeng's Annan (present-day Qinlong) under its jurisdiction, there were overlapping valleys extending 10,000 li, with towering old trees. According to Shuichengting Caifang (Interviews in Shuicheng Prefecture) in the reign of Guangxu Emperor, in west Guizhou there were dense forests that extended for more than 100 li in length and breadth. In southeast Guizhou, according to Qiannan Zhilüe (Brief Annals of Southern Guizhou) by Guizhou governor Ai Bida in the fourteenth year in the reign of Qianlong Emperor, 'along 200 li of the river banks from Qingjiang (present-day Jianhe County) to Maoping (present-day Jinping County), there are clouds sheltering from the sun, and no land is exposed. There are no places that receive no shade of trees'.

The 1989–1993 figures of the fourth national forests resources inventory show that the forest coverage rate in Guizhou was only 14.75% (Editorial Board of *Forests in China* 1995: 231). Apart from Qiandongnan Prefecture and Chishui City that had a higher rate of forest coverage, the rate in most parts of Guizhou was lower than ten percent, and the virgin forestlands had been reduced to almost nothing. The rapid deforestation had given rise to serious soil erosion in Guizhou's karst areas. In 1980, there was soil erosion in Guizhou, and rocks were exposed in larger areas. The area of degraded land amounted to 134,000 km^2, which consequently increased to 23,500 km^2 in 2003. Each year the silt carried from Guizhou into the Yangtze and the Pearl rivers amounted to about 0.2 billion tons (Liu 2003), which not only brought enormous destruction to local social and economic development, but also presented threats to the ecological security of the lower reaches of the Yangtze River. The deterioration of the ecological environment also had brought over 500,000 farmers into absolute poverty. Although the government has offered support to them for many years, they still do not have enough food to eat or clothes to wear and their homeland can no longer support them.

Guangdong

The mountainous karst areas are connected with their counterparts in Guangxi, under the jurisdiction of Qingyuan and Shaoguan cities. In the Qing dynasty, the region was covered by dense forests, 'Most of the trees are *tu* trees, *ranmu*, pine, cypress, thorny trees, mulberry, *yung*' (Wang 1673). By the early 1980s, the region had witnessed the serious deterioration of the ecological environment, becoming one of China's top ten poverty-stricken areas at that time, and agricultural production had failed to support the local people to alleviate their poverty.

Ecological changes in northeast Yunnan and Nujiang River Valley

In the Ming and Qing dynasties, northeast Yunnan was the region for producing the *namu* trees that were to be used in building the large-scale imperial architectural projects. In the reign of Guangxu Emperor of the Qing dynasty, this region was 'covered by dense bamboo groves and big trees, with venomous snakes and fierce beasts as well as green bushes and cold wind, and nobody dared to access it' (Wu and Song 1887). By the early Republic of China, 'there were very old forests, with big trees lying on the ground like old dragons, and their foliage giving shade and shelter from the sun, which makes the day as the dark night with many wild animals in the forests' (B. Chen 1919). However, there had been large-scale deforestation along the main streams and tributaries of the Jinsha River. In the 1950s, the forest coverage rate in northeast Yunnan was about 19.8%, which was reduced to 6.3% by the 1980s (Lan 1992: 65).

In the 1950s, the forest coverage rate was more than fifty percent in Nujiang River valley. After thirty years of deforestation, the rate was reduced to 33.6% in the 1980s (Policy Research Section, et al. 1998: 15). In the 1990s, no forests were to be found in the lower half of the river valley below an altitude of 2000 meters. The dramatic reduction of forests in the above-mentioned two areas has heavily destroyed their ecological environment, producing a lot of victims of ecological disasters.

Causes of ecological change

Population pressure and ecology

Population pressure is one of the gravest problems faced by contemporary people and is also the source and core of all secondary environmental problems. The ecological changes in south China's karst and high, cold mountainous areas are also related to population pressure.

The Chinese government has not implemented a strict family planning program in border and minority areas, while the medical treatment has improved greatly. Therefore, in the 1990s, south China's karst and high, cold mountainous areas saw a rapid population growth. In 1992, the population density of ten counties and cities of Mashan, Tiandeng, Pingguo, Jingxi, Xincheng, Du'an, Yizhou, Debao, Daxin and Hechi in Guangxi's karst and high cold mountains was higher than China's average population density of 121 persons. Guizhou, where karst areas form the major topographical features, had a population of a mere 14.74 million in 1949, which had increased to 35.245 million by the end of 2000, a total increase of 20.505 million in a matter of fifty years. Usually it is thought that in order to be accommodated by its environment Guizhou's population should be less than twenty million. That is to say, the present human carrying capacity of Guizhou has surpassed forty-three percent of its ideal capacity. There has also been surprising population growth in northeast Yunnan and in Nujiang River valley. In 1950, Nujiang Prefecture had a population of 203,500, and in

1990 when the fourth population census was carried out, its population was 434,800 (Editorial Board of *Kuashiji de Zhongguo Renkou* 1994: 26). Over forty years, the population had grown by more than twice as much. Under the pressure of population explosion, and in order to find a solution to the basic human need for food, shelter and firewood, and when the production forces and modes of production had not yet been obviously promoted and transformed, the minority people in the above-mentioned two regions had to live by reclaiming wasteland and deforestation, resulting in the destruction of their ecology.

Restrictions on population movement and regional ecological deterioration

China has a long history of compulsory household registration, but such a system that exercises control over population movement was implemented mostly in areas where the formal prefectures and counties were established, while border and ethnic minority areas were basically free of such surveillance. However, since the 1950s, things have changed. China's powerful state organization had made full use of this age-old household management system. On January 9, 1958, the Standing Committee of China's National People's Congress convened its ninety-first session, and adopted *Regulations of the People's Republic of China on Household Registration*. Since then, this new household management system has been launched nationwide. This regulation stipulates that 'those who are to migrate from rural to urban areas are expected to hold employment certificates issued by urban labor departments, admission certificates issued by schools or certificates that are issued by the urban household registration organs and allow them to move'. Each urban household holds one residence booklet, while rural households were issued such booklets with the cooperative as a unit. The residence booklet had the same validity as the ID card, and because the farmers did not have such a booklet, they had no certificates of identification with them when going out of their villages. Although the household registration management system served to restrict the movement of rural populations into urban areas, the farmers, under the planned economic system and the people's commune management system, had to ask for leave with certificates issued by the production brigades, without which they were not allowed to stay in hotels and not allowed to buy tickets for vehicle or boat transportation. The food coupon was required when dining out, and only the non-rural population was issued such coupons. Thus, the household registration system actually also restricted the farmers' rural movements. With this restriction, the surplus labor could not be absorbed by urban areas, and people were not allowed to move freely for agricultural purposes. For example, in the 1970s and 1980s, several thousands of the Miao people moved spontaneously from northeast Yunnan and northwest Guizhou, and reclaimed wasteland and took up farming in the mountainous areas of Anning City near Kunming, the capital of Yunnan Province. In response, their newly-built houses were burned down and they were sent by force back to their original villages. The restriction of free movement has resulted in low productivity and the increased population has placed further pressure on the ecology of the karst areas and high, cold mountains.

Impact of agricultural policies on the forest environment

Before 1980, China long promoted a grain-focused agricultural policy called 'taking grain as the key link', which produced large areas of deforestation in the name of opening up uncultivated 'wasteland'. For example, in the Great Cultural Revolution (1966–1976), the agricultural policy of taking grain as the key link and felling all trees was executed in Nujiang Prefecture, which destroyed a lot of forest reserves and finally led to the disastrous mud-rock flows in large areas in the prefecture in 1979. This disaster destroyed 61,928 *mu* of farmland, 2193 canals, forty-one bridges and culverts, and 2047 houses (Office for Anniversary Celebrations 1984: 107).

Process of ecological migration

Migration in the karst mountainous areas

Guangxi

Since 1993, the Guangxi government has decided as part of their poverty-relief program[3] to move needy families whose per capita cultivated land is less than 0.3 *mu* in the rocky mountains and to settle them in other areas where there is convenient availability of land, water and transportation, and help them build new homes and find a solution to food and clothing issues. After ten years of ecological migration, that is, by the end of December 2002, more than 50,000 households of about 230,000 needy people have been moved from their original rocky mountainous homelands. The government has built 267 farms and 1167 settlements for them. Among these 230,000 migrants, those from Hechi and Bose accounted for half each. One-hundred-and-twenty thousand people of the Miao, Yao, and Zhuang moved from Hechi area, and built seventy-four settlements and 494 residential areas (*Xinhua Tongxunshe* 2003). A total of 110,000 needy people of the Miao, Yao, and Zhuang were moved from around Bose. According to the media, the migrants achieved the goals of entering the new areas, residing in new houses, developing the area and having enough food and clothing all in the same year (T.H. Chen 2000: 12).

Guizhou

Since 1986, Guizhou provincial government has adopted migration as a measure of assistance to particularly needy populations, through moving and settling them in ecologically better areas. By 2001, a total of 17,817 households of 852.370 million people had been resettled (Ran 2001: 47). The migration has relieved the ecological pressure once exerted by the villages, and the new settlers have been offered much better conditions of production and daily life, and have enough food and clothing. However, owing to the very large area of ecologically deteriorated terrain in Guizhou, there are still 460,000 people to be resettled for ecological reasons. Guizhou Province plans to move these people in the next ten years in order to help them throw off poverty once and for all.

Guangdong

In 1991, officials from the central and local governments went on an inspection tour of Qingyuan City in order to understand the living conditions in marginal and poor areas. Owing to the very hard, tough and dangerous road, their vehicle had an accident that almost resulted in their death. The event reminded the officials in charge of poverty alleviation that the karst mountainous areas such as Qingyuan involve long-distances and dangerous transportation facilities, and cannot be expected to introduce significant outside enterprises in its economic development. Even though there are some enterprises doing business in the area, it is very hard for their factories to sell their products to the outside world. Businesses must venture outside the region to conduct trade in order to relieve poverty in the mountainous areas. Therefore, the Guangdong provincial government decided that industrial development zones should be established near Qingyuan and Guangzhou, and that needy farmers should be moved to this location from the mountainous areas. In the three years from 1993 to 1995, a total of 100,000 people in Qingyuan were moved from the rocky mountains. From 1993 to 1998, a total of 7414 households of 34,072 people were moved from Shaoguan City. The two cities moved a total of over 130,000 people. The migrants experienced a significant increase of income and thus had sufficient food and clothing in a short period of time.

Migration in northeast Yunnan and Nujiang River valley

In the 1990s, there were more than 500,000 people in Yunnan who had lost their basic living conditions and needed to be moved to other locales. Of these areas, northeast Yunnan and Nujiang River valley were the most concentrated areas where local people had lost their basic living conditions.

In the 1990s, there were more than 400,000 people in Zhaotong Prefecture who were especially poor. Of these people, 200,000 had basically lost their living conditions, most of whom were Miao people from high and cold mountainous areas. In Nujiang Prefecture, there were 54,000 Lisu and Nu peoples who were in the desperate condition of having no access to water, land, forests and roads, and they were badly in need of migration for ecological reasons. In order to solve the problem of survival among these victims of ecological disasters, since 1996 Yunnan Provincial Government has initiated a pilot migration project. The migrants should meet two requirements, where: 1) the ecological environment of their residential area is poor and frequently affected by natural disasters, the local people have basically lost their living conditions placing a heavier burden on the land, and the per capita cultivated land area is less than 0.3 *mu*; and 2) they are able-bodied (no more than forty years old in principle), abide by the laws and regulations and follow the family planning program. By December 2003, Yunnan had organized the migration of 190,000 people to Simao City and Jiangcheng County in its southwest part. Most migration is concentrated in Dehong Prefecture in west Yunnan. Migrants from northeast Yunnan settled in Simao and those from Nujiang River valley settled in Dehong. All the migrants basically met the government's expected target of migrating to the outside world, settling down and growing rich.

According to incomplete statistics, by the end of 2003, a total of 640,000 most needy people had been moved from the karst and high, cold mountainous areas. Ecological migration has been completed for the time being in Guangxi and Guangdong, while there remain 310,000 and 460,000 people to be moved in Yunnan and Guizhou respectively. If the grain for green[4] program is to be executed strictly in southwest China, an additional 360,000 ecological migrants will be required in Yunnan alone.

Patterns of ecological migration

Northwest China's arid and semi-arid areas are the areas where the Chinese government launched the poverty alleviation by ecological migration strategy for the first time, and the scale of such migration far surpassed its counterparts in southern China's karst and high, cold mountainous areas. According to incomplete figures, in the 1990s, 540,000 people were moved in Gansu Province (Luo 2000: 201) and 1,000,000 in Ningxia Hui Autonomous Region (Lü 2000:5). Three patterns of migration may be distinguished: 1) Individual migration—poor families spontaneously seeking refuge with their relatives and friends, or scattered settlers with certain government allowances; 2) Migrant settlement in the development bases sponsored by the government; and 3) Migration with two home villages—these migrants have homes both in their former village and in their new settlement, and completely move to the latter when the development, production and living conditions are stabilized in the new settlement. These patterns actually fall into two categories: collective and scattered migration. Seen from the migrants' perspective, it is still agriculturally motivated migration.

The governmental patterns of migration in south China vary according to the amount of usable land in the settlement, as well as the different levels of economic development and financial income in different provinces and autonomous regions.

Long- and short-distance migration

According to the distances involved in migration, there are two kinds of ecological migration in south China. The first is long-distance migration that crosses prefectures or cities, which is the case for most of the migrants in Guangdong and half of the migrants in Yunnan and Guangxi. This pattern of migration is the most costly, and requires a level of economic aid provided by the central and provincial governments to the local governments and residents. Yunnan's per capita settlement allowance is more than 15,000 *yuan*. The administrative coordination is more difficult, that is, the recipient governments and residents are usually opposed to mass migration from outside. They will not actively receive any of these migrants unless their superior governments intervene. However, because of the significant distances and the better conditions at the recipient places, long-distance migration has a lower rate of migrants returning to their original villages.

The second is short-distance migration that takes place within one county or one township, which is the case for Guizhou's migrants. This kind of migration has a lower cost, and the per capita settlement allowance is usually about 5000 *yuan*. Administrative coordination is easier, the settlement is usually a smooth process and the culture of the migrant groups can be maintained for a longer period of time. The disadvantage with short-distance migration is that the migrants' economic development is slower.

Migration by different means of subsistence

Viewed from the perspective of migrants' subsistence, south China's ecological migration can be divided into the following kinds.

Urbanized migration

Urbanized migration is that in which the migrants completely change their line of activity from rural to urban industries. Guangdong Province embraced industrial migration policies wherein they moved the poor population from the northwest of the province to the industrial development zones near major cities such as Guangzhou or Qingyuan, so the settlers were completely transformed from farmers into workers. This kind of migration needs more financial support and industrial absorption, to employ the new settlers. Guangdong was one of China's earliest areas to open up to the outside world and thus is the most developed economically. In the 1990s, Guangdong was badly in need of labor for its economic development, and it was easy to provide jobs for 100,000 to 200,000 people. Guangxi is located at the juncture of central and west China, with slower industrial development than Guangdong but faster than that in more westerly Yunnan and Guizhou. Nearly 100,000 of its 230,000 migrants were settled in the industrial development zone and the tertiary industry of the coastal Beihai City. The advantage of urbanized migration is that it helps a lot of farmers out of poverty in a very short period of time and requires no land reclamation for agriculture, so no additional ecological destruction is expected. This is ecological migration in its true sense. However, the disadvantage of urbanized migration is that the culture of the migrants is doomed to disappear soon after resettlement takes place.

Migration with the integration of agriculture and industry

This pattern is often called the 'company plus farmers' migration strategy. Its distinguishing features are: 1) that the agricultural development zone is established in places where the land is connected as a single unit and is of larger size with better conditions of sunshine, heat, water and soil; 2) that it occurs with the precondition that farmers' food-production land is guaranteed; and 3) that the economic development is carried out in a cooperative form between the company and farmers. This is the case in Yunnan's cross-prefectural migration and a portion of Guangxi's migration. The government invests in the company in establishing a series of plantations of coffee, tea, rubber or sugarcane, and the company organizes the migrants in producing commercial products. This migration pattern not only takes into account the farmer's complex relationship towards land and agriculture, but is also not a simple repetition of their former subsistence means. This cooperative form of migration is thus a good option theoretically and practically. However, the company's business runs some market risk, which demands a certain degree of psychological enduring capacity from the farmers. If the number of immigrants is too large, this pattern of farm type will also result in ecological degradation in the future.

Agricultural migration

This pattern requires non-stop agricultural production activities in the geographical relocation, which has been the case with migration in the provinces and autonomous regions of south China. Farmer's acquisition of enough land in the new settlements is the basic guarantee for the success of such a migration pattern. The advantage of agricultural migration is that such migration is less costly and easier to manage than other forms. In its execution, it is further divided into collective migration and individual settlement. The latter requires migrants to settle down individually in some new settlements. However, this kind of migration will increase the pressure on the ecology of the new settlement.

Conclusion

Ecological migration is the best way of solving the problem of survival among the poor population as well as the ecological crisis in the karst and high, cold mountainous areas in south China. Ecological migration that aims at poverty alleviation is actually an organized, planned and policy-based migration strategy. Seen from the migrant's social and cultural adaptation, spontaneous migrants are well prepared before migration, so they have a stronger sense of responsibility for their own behavior, and this responsibility can help them overcome various problems and more easily adapt themselves to their new environment or community. However, organized migration is migration under the government's unified leadership, and these migrants psychologically rely on the government and its policies, presenting more difficulties in settling them. The ecological and social effects of large-scale organized migration will be measured over many years to come. Viewed from past experiences, large-scale organized migration makes a greater disturbance in the ecological environment of new settlements. The 'company plus farmers' migration strategy and agricultural migration are not counted as ecological migration, for they present a potential threat to the natural ecology of the new settlement, so it is more appropriate to label these forms 'environmental migration'. At present, west China's natural conditions are very adverse. People in many areas are not guaranteed basic living conditions, and there are over five million people waiting to be resettled (*Jingji Ribao* 2001). Migration of such magnitude requires extensive governmental financial support. If migration is not carried out properly, it will bring additional ecological disasters to other parts of west China. According to the latest official figures, the forest coverage rate of west China is only one third of that for east China, and what is more, east China is economically developed and its urban areas have a larger capacity to absorb additional population. I propose that west China's ecological migration should follow the migration pattern of the Three Gorges Dam in the Yangtze River, that is, to migrate from west to east, especially to the coastal areas in southeast China. In March 2004, at the second session of China's tenth National People's Congress, Mr. Chen Derong, the deputy to this congress and the mayor of Jiaxing City of Zhejiang Province, came up with the same idea. He said, 'the state should encourage migration from west to east China'. He continued to explain,

> At present, the industrialization and urbanization are gaining full speed, but their population is increasing at a lower speed, and they are badly in need of additional labor. However, the central and west parts of China are experiencing employment pressure, and their ecological environment is rather fragile. Therefore, taking the country as an entity, part of west China's population should be moved, through market mechanisms, to the east part of China. The state should launch the policies encouraging east China to absorb the population from central and west China. In order to urge such a migration, it is suggested that the central Chinese government should increase, with the urban planning standards of 8000 to 100,000 people per km^2, the urban construction land for the migration in east China's cities so as to expand more development room for the east areas. (*Renmin Ribao* March 9, 2003)

Notes

1. Poverty is defined as the case in which the living standards of one person or a family fails to meet the minimum standards accepted by a society.
2. Absolute poverty or alternatively poverty of survival refers to the fact that under a given mode of production and life style, one person or one family's income from work and other lawfully earned income cannot meet the basic needs of survival, and such person or family is referred to as poverty-stricken. In this case, and in terms of production, the labor force so lack the material conditions for reproduction that they find it hard to maintain their own simple reproduction, and the producers can only reproduce at a reduced scale. In terms of consumption, such people do not meet the most basic human needs of survival in clothing, food and shelter, etc.

3. The Chinese government's poverty alleviation program has been executed in three periods: 1) 1978–1985, the period of poverty alleviation by institutional restructuring; 2) 1986–1993, the period of full-scale poverty alleviation; and 3) 1994–2000, the period of storming the stronghold of poverty, or of reducing the number of needy people by eighty million in seven years.
4. This program, launched with a decree from the central Chinese government issued in the wake of the flood in the lower reaches of the Yangtze River in 1998, requires cultivated slopes over twenty-five degrees in the upper reaches of the Yangtze River to be returned from grain to green, and no land reclamation is allowed.

References

Chen, B. (1919), *Wuchan* (Products), Volume 6 of Zhaotong Deng Baxian Tushuo (Pictorial account of Zhaotong and seven other counties), letterpress printed by Yunnan Society in the Republic of China.

Chen, T. (2000), 'Guangxi Bose Diqu Jianchi Jiudikaifa yu Yidikaifa Xiangjiehe', (Guangxi's bose area adheres to the integration of local and non-local development), *Renmin Ribao* (People's Daily), January 13.

Editorial Board of Forests in China (ed.) (1995), *Zhongguo Senlin* (Forests in China), Volume 1, Beijing: China Forestry Publishing House.

Editorial Board of Kuashiji de Zhongguo Renkou (China's population at the turn of the century) (1994), 'Yunnan Juan' (Volume of Yunnan), in *Kuashiji de Zhongguo Renkou* (China's population at the turn of the century), Beijing: China Statistics Publishing House.

Jingji Ribao (Economic Daily) (2001), 'Zhongguo Nongcun Fupin Kaifa Baipishu' (White paper on China's poverty alleviation and development in rural areas), October 16.

Lan,Y. (1992), *Lishishiqi Xi'nan Jingjikaifa yu Shengtaibianqian* (Southwest China's economic development and ecological changes in historical periods), Kunming: Yunnan Educational Press.

Liu, L. (2003), 'Guizhou Kasite Shimohua Qushi Yanjun Yuanshi Zhuanjia Tichu Bixu Jiakuai Zonghe Zhili' (The desertification trend is serious in Guizhou's karst areas, and the academicians and experts suggest it should be tackled in a comprehensive way), *Guangming Ribao* (Guangming Daily), January 13.

Luo, G. (ed.) (2000), *Zhongguo Caizheng Fupin Wenti Yanjiu* (A study of the problem in China's poverty alleviation by providing financial input), Beijing: China Financial and Economic Press.

Lv, A. (2000), *Guonei Kaifa Yimin Zhiduoshao* (How many domestic migrants of a development kind), *Jingji Ribao* (Economic Daily), August 22.

Office for Anniversary Celebrations of Nujiang Prefecture (ed.) (1984), Qianjin zhong de Nujiang (Nujiang is forging ahead), printed for internal use.

Policy Research Section of Yunnan Provincial Committee of the Communist Party of China, and the Compiling Committee of Yunnan Provincial Local Chronicles (eds) (1988), 'Yunnan Dizhoushixian Gaikuang-Nujiang Lisuzu Zizhizhou Fence' (Book of Nujiang Lisu Autonomous Prefecture), in *A Survey of Yunnan's Prefectures, Autonomous Prefectures, and Counties*, Kunming: Yunnan People's Publishing House.

Ran, M. (2001), Yimin Banqian shi Jiejue Tekun Renkou Wenbao Wenti de Youxiao Tujing (Migration and resettlement is an effective way of solving the problem of food and clothing among the most needy people), *Guiyang: Guizhou Minzu Yanjiu* (Ethnic Studies in Guizhou), 2.

Renmin Wang (People's Daily Online) (2003), *www.people.com.cn*, March 9.

Wang, Y. (1673), *Xinxiu Guangzhou Fuzhi* (Newly Compiled Annals of Guangzhou Prefecture), hand-copied book in the reign of Kangxi Emperor of the Qing dynasty, Volume 10

Wu, G. and C. Song (1887), Yiwen (Art and literature), Volume 6 of *Zhenxiong Zhouzhi* (Annals of Zhengxiong Prefecture), block-printed edition in the reign of Guangxu Emperor of the Qing dynasty.

Xiao, Y. (1996), *Guangxi Renkou yu Jingji Ziyuan Huanjing* (Guangxi's population and its economic resources and environment), Nanning: Guangxi People's Publishing House.

Xinhua Wang (Xinhua Online) (2003), 'Hechishi Shi'erwan Nongmin Zouchu Dashan' (120,000 farmers migrated from large rocky mountains in Hechi City), Guangxi (*www.gx xinhuanet.com*), September 30.

Yang, J. (ed.) (1998), *Guangxi Quqing Jiaoyu Duben* (A reader in learning Guangxi's regional conditions), Nanning: Guangxi People's Publishing House.

Zhu, H. (1985), 'School notes of Xu XiaKe Travels', Yunnan People's Publishing House, translated from Chinese by Chen Jianming.

5
Reforestation after Deforestation in China

Liu Dachang[1]

Introduction

Over a long period many factors have combined to result in deforestation and serious forest degradation in China. These include the preparation of land for farming and settlement, the provision of fuel for heating, cooking and industrial processes and the supply of timber for activities such as construction. Forest cover was already limited when the People's Republic of China was founded in 1949. Since then there have been significant changes in China's development and forest policies. Some policies unintentionally caused over-logging, and on three occasions from the 1950s to the 1980s periods of destructive deforestation occurred. During the 1990s, especially the late 1990s, forest policies emphasised environmental protection and ecological benefits. The government expanded existing environment-focused forest programs and launched additional projects. At the same time there were government initiatives to develop plantations primarily for wood production to meet China's growing need for construction timber and paper.

The over-logging, deforestation and forest programs had, and continue to have, enormous environmental impacts. This chapter describes these programs and discusses their environmental impacts. It also briefly analyses the causes of the deforestation and over-logging, and suggests how the forest programs can provide environmental benefits and employment and income generation.

Deforestation and over-harvesting from the 1950s to the 1980s

Large-scale deforestation took place nationwide in extensive collective forests[2] in the 1950s, 1960–70s and 1980s, while over-logging occurred in both state and collective forests. These events provided many lessons concerning their causes and environmental impacts.

Three nationwide deforestation cases

The first nationwide deforestation took place following the Chinese Government's radical 'Great Leap Forward' campaign in 1958 to intensify the country's agricultural collectivization through the people's communes and to boost China's steel output to a level that would surpass that of Great Britain in fifteen years. People were mobilised throughout China and forced to make iron and steel using primitive 'backyard' furnaces. About 600,000 of these furnaces were in operation by October 1958 (Clubb 1972). This industrialization policy caused significant loss of forest and biodiversity because energy for the furnaces was provided mainly by wood and charcoal. For example, in only one village in Cangwu County, Guangxi, 190 charcoal-making sites were established. Their wood was supplied by clearing patches of high quality evergreen broadleaf forest and harvesting in other dense forests resulting in poor quality low-stocked secondary forest (Liu et al. 2003). Agricultural collectivization contributed to large-scale deforestation because it led to a radical transformation from private to collective ownership (Menzies and Peluso 1991).

Three factors are believed to have contributed to the second serious deforestation event which took place in the 1960s and 1970s. The 'Great Cultural Revolution' campaign, initiated in 1966 and continuing for ten years, paralysed almost the whole governance system from the central to village level and resulted in the breakdown of the management system of collective forests in many rural areas. Village leaders lost control over forest management, and illegal logging began and increased in many places (Liu et al. 2003). Secondly, development policy at that time set grain food production as an essential component of national economic development and over-emphasised local self-sufficiency in food. This policy caused large areas of forest and grassland to be cleared and cultivated for food production. Thirdly, shifting cultivation/swidden agriculture, traditionally practiced by some minority groups, became unsustainable because increasing population pressure on land resulted in much shorter fallow periods. This factor contributed especially to forest loss in the southwest provinces of Yunnan and Sichuan.

The third and most disastrous nationwide period of deforestation occurred in the early- and mid-1980s, immediately after the forest management reform. In 1978, China began agricultural reform that abolished collective agriculture and introduced household-based production. Use rights to farmlands were distributed among individual households, while the village collectives remained the land owners. Along with the transfer of land-use rights, households gained increased autonomy to make decisions about land use, agricultural production and product marketing. Forest management reform in the early 1980s largely followed the pattern of the agricultural reform. Use rights to denuded or non-forested collective land were distributed to farm households as their family plots in an attempt to encourage them to plant trees and develop plantations. At the same time, management responsibility for existing collective forests was transferred to farm households in a bid to improve their management. However, the results of the forest reform were not what the policy makers had planned or anticipated. Instead of planting trees and improving forest management, significant deforestation occurred in most villages, especially in south and southwest China where collective forests accounted for sixty to ninety percent of the total forest area. Some local governments responded to this serious deforestation by converting from household to collective management. Most forest management research indicates that forest tenure insecurity, due to frequent tenure changes from the 1950s to the 1970s, was the primary contributing factor to the deforestation (Yin 1994; Liu 2001; Liu and Edmunds 2003a; 2003b; 2003).

Over-logging

In addition to the deforestation events, over-logging also occurred in both state-owned and collective forests. Nationally, state-owned forests represent about forty percent of China's total forest area. They are largely located in the north eastern provinces of Heilongjiang, Jilin and Inner Mongolia and the south western provinces of Sichuan and Yunnan. In these two regions, from the 1950s to the 1970s, the government established over 130 large state-owned logging enterprises to meet the great need for timber for industrial development. Large-scale logging in southwest China did not end until the pilot phase of the Natural Forest Protection Program (NFPP) in the very late 1990s, while logging in the northeast is still continuing, though at a considerably reduced level after the implementation of the NFPP. To complement timber harvested from state forests, significant quantities of timber were cut from the collective forests in south and southwest China, where the natural vegetation is evergreen broadleaf forest with rich biodiversity.

Environmental impacts

Four decades of deforestation and over-logging have had environmental impacts including: forest loss and degradation, biodiversity loss, decreased eco-services and further damage to forests, other vegetation cover and ecosystems. According to national forest inventory (NFI) figures, the country's total growing stock dropped from 9.57 billion m^3 in the 1977–81 NFI period to 9.42 billion m^3 in the 1984–88 NFI period.[3] Total forest area decreased from 121.9 million ha in 1973–76 to 115.3 million ha in 1978–81. Although forest area increased from 1977–81 to the 1984–88 period, as a result of the deforestation and over-logging there was less natural forest and more artificial plantations (SFA 2005). The ratio of plantations to total forest area increased from 19.9% in 1977–81 to 24.7% in the 1984–88 period (MoF 1989). Natural forests are complex ecosystems, often with high levels of biodiversity; harvesting reduces or eliminates their biodiversity. Forest plantations are often monocultures, or include only a few species, and so are able to provide less eco-services including control of soil and water erosion.

Reforestation and forest protection since the 1990s

In response to ecological and environmental degradation, the Chinese government changed its forest policy from a focus on timber/wood production to a focus on environment. By the late 1980s, the government had formulated a comprehensive plan for increasing national forest cover. The plan is a product of the recognition of the need to use forests as an effective means of rehabilitating damaged ecosystems, while meeting increasing domestic demand for wood and other forest products. This policy change 'coincided' with the call by the Rio Declaration on Environment and Development, Agenda 21, for sustainable development.

More frequent and serious natural disasters in the 1990s, such as floods and sand storms, and especially disastrous floods in 1997, alerted the government and people to the fragility of China's environment. This recognition, together with an improvement in the ecological and environmental awareness of the Chinese people, brought about a further evolution of China's forest and environment policy in the context of sustainable development. Three new environment-focused forest programs have been initiated, while those existing forest programs have been repackaged into three programs with expanded scope and investment. Out of these initiatives, five are reforestation-oriented or have a reforestation component. These five forest programs and some international aid reforestation projects and their environmental impacts are discussed below. Their profiles are shown below in Table 5.1.

Natural Forest Protection Program

The Natural Forest Protection Program (NFPP) was officially launched in October 2000, after a two-year pilot phase. It is a ten-year program, implemented from 2001–2010, involving a planned investment of 96.2

Table 5.1 China's six mega forest programs

Programs		Duration	Location and Scope	Goals, Targets and Estimated Investment (*Yuan*)	Progress and Achievements
Natural Forest Protection Program		2001 – 2010 (pilot phase: Oct 1998 – Sept 2000)	Upper Yangtze River Watershed, Mid-to-upper Yellow River Watershed, Xinjiang and Hainan, Northeast China State Forest Region	• Protect existing natural forests and develop plantations to restore cegraded environment • Over 100 billion	3.5 million ha of plantations established and 4.1 million ha of secondary forests naturally regenerated from 2001 – 2005
Farmland-to-Forest Program		2000 – (pilot phase: 1999)	Anhui, Beijing, Chongqing, Gansu, Guangxi, Guizhou, Hainan, Hebei, Heilongjiang, Henan, Hubei, Hunan, Inner Mongolia, Jiangxi, Jilin, Liaoning, Ningxia, Qinghai, Shaanxi, Shanxi, Sichuan, Tianjin, Tibet, Xinjiang, Yunnan	Establish 32 million ha of tree plantations	• 103 billion spent from 1999 to 2005 • 21.6 million ha of tree plantations established by end of 2005
Combating Desertification Program		2002 – 2010 (pilot phase: 2000 – 2001, fully launched in 2002)	Beijing, Tianjin, Hebei, Inner Mogolia and Shanxi	• Plant 7.57 million ha of trees and restore 10.6 million ha of grasslands • 55.9 billion	2.6 million ha of tree plantations established from 2001 – 2005
Shelterbelt and Windbreak System Program	Three-North Shelterbelt	1978 – 2050	Beijing, Gansu, Hebei, Heilongjiang, Inner Mongolia, Jilin, Liaoning, Ningxia, Qinghai, Shaanxi, Shanxi, Tainjin, Xinjiang	Establish tree plantations as shelterbelt to prevent and control desertification and restore desertification lands in the Three North Shelterbelt	By 2005 • 3.97 billion *Yuan* spent • 18.0 million ha of tree plantations established and 7.1 million ha of natural regeneration forests
	Mid-Upper Yangtze River Watershed Conservation Forests	1989 –	Anhui, Chongqing, Gansu, Guizhou, Henan, Hubei, Hunan, Jiangxi, Qinghai, Shaanxi, Sichuan, Yunnan	Improve tending existing middle and young natural forests and tree plantations and establishing new tree plantations to protect headwaters and rehabilitate habitat in the upper and middle reaches of the Yangtze River	1.44 million ha of plantations established during the period 2001 – 2005 (including Pearl River Watershed Conservation Forests, Coastline Windbreak Forests, Plains Greening / Agroforesrty, and Taihang Mountain Greening)
	Pearl River Watershed Conservation Forests	1995 –			
	Coastline Windbreak Forests	1991 –	Anhui, Fujian, Guangdong, Guangxi, Hainan, Hebei, Jiangsu, Liaoning, Shandong, Shanghai, Tianjin, Zhejiang		
	Plains Greening / Agroforesrty	1987 –	Anhui, Beijing, Fujian, Heilongjiang, Hebei, Henan, Hunan, Jiangsu, Liaoning, Shandong, Shanghai, Shaanxi, Shanxi, Sichuan, Zhejiang		
	Taihang Mountain Greening	1994 –	Beijing, Hebei, Henan, Shanxi		
Reforestation for Wood Production Program					357,000 ha of fast-growing, short-rotation plantations established during 2001 – 2005
Nature Reserves and Wildlife and Plants Conservation Program		2001	Throughout the country	Establish and manage nature reserves totaling 155 million ha in 2010 and 16[illegible] million ha in 2030	

Figure 5.1 NFPP and Combating Desertification Program

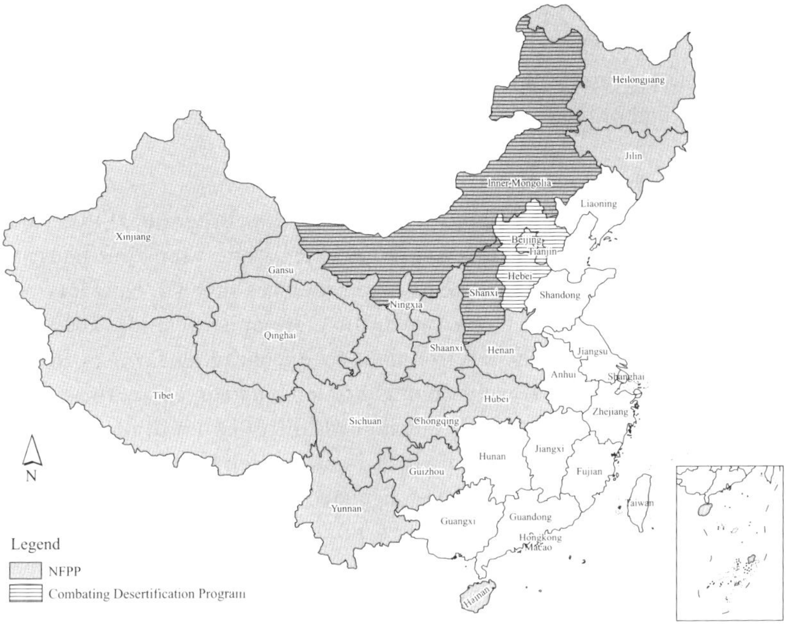

billion Chinese *yuan* from the central government, or more than 100 billion *yuan* if the costs of the pilot phase are included (*China Green Times* 2003). The NFPP was designed to conserve the existing forests by: 1) enforcing complete logging bans in the upper Yangtze River watershed (Chongqing, Guizhou, Hubei, Sichuan, Tibet, Yunnan), mid-to-upper Yellow River watershed (Gansu, Henan, Inner Mongolia, Ningxia, Qinghai, Shaanxi, Shanxi), Xinjiang and Hainan; and 2) reducing logging in the northeast China State-owned Forest Region (Jilin, Heilongjiang) as well as increasing forest area by plantation establishment and natural regeneration to restore the degraded environment in the project area (Figure 5.1). To achieve these goals, the duties of more than half a million staff of the state logging enterprises affected by this program were changed from logging to forest tending and patrol duties.

In its first five years (2001–2005), the NFPP advanced toward achieving its objective of ecological restoration. Forest depletion was effectively controlled, and forest area and growing stock increased. Commercial timber harvesting virtually stopped in the upper Yangtze River and mid-to-upper Yellow River. Annual timber output in the northeast China State-owned Forest Region was reduced by nearly forty percent, from 18.2 million m^3

in 1997 to 11.2 million m^3 in 2005. Some 3.5 million ha of plantations were established and 4.1 million ha of secondary forests naturally regenerated through mountain land closure (*China Green Times* 2006a). This progress indicates the success of this program in effectively conserving forests and biodiversity and providing ecological benefits.

Farmland-to-forest program

The 'Sloping Lands Conversion (*tuigeng huanglin)* Program' (SLCP) was launched in seventeen provinces in 2000 after a one-year pilot phase in 1999 and expanded to include twenty-five provinces in 2002. Now it covers 2279 counties and involves thirty-two million farmer households. The SLCP was initiated to restore the environment by encouraging farmers to stop cultivating food crops on steeply sloping farmlands and to replace these crops with tree plantations. It planned to establish thirty-two million ha of tree plantations by the end of 2010, of which 14.66 million ha will be developed on farmlands and 17.34 million ha on denuded forestlands.

To encourage conversion of the farmlands, the central government provides farmers participating in the program with funds in the form of food grain and cash. Farmers planting trees on one hectare of farmland receive 1.5 tonnes of food grain a year in the upper and middle reaches of the Yellow River, and those in the upper Yangtze River watershed will get 3.0 tonnes. Participating farmers will receive the food compensation for five years if they plant 'cash trees' yielding non-wood forest products (NWFP), and eight years if they establish tree plantations aimed at providing ecological benefits. In addition, participating farmers will get cash compensation of 300 *yuan* per ha per year, and cash subsidies of 750 *yuan* per ha to buy tree seedlings or grass seeds at the beginning of their participation in the SLCP. The funding policy has recently been changed to allow a longer period of compensation for participating farmers to conserve the established plantations. Compensation for an additional eight years is given to farmers who have developed tree plantations focused on providing ecological benefits, and an extension of five years is given to those who have planted NWFP trees. This is necessary because if compensation is stopped, there is a high probability that participating farmers will cut down the established plantations and convert the area to food production.

By the end of 2005, 21.6 million ha of tree plantations had been established (nine million ha on slope farmlands and 12.6 million ha on denuded forestlands) under the SLCP. These plantations represent about sixty-seven percent of the program's target. This progress is largely attributed to the financial incentives introduced by the central government that provided 103 billion *yuan* for the program from 1999 to 2005 (*China Green Times* 2007a). Farmers' interest is also a contributing factor. Participating farmers benefit from the cash and food grain compensation that represents an immediate incentive for them as the amount of grain they receive is more than the output they normally obtain from their poor quality land. A long-term benefit for the farmers is expected to be an improvement in the local climate due to tree planting.

The timber plantations will certainly have positive environmental impacts, including soil erosion control, carbon sequestration and other ecological benefits. Greater environmental impacts may be achieved if the SLCP can find a way to encourage farmers to develop multi-species plantations to provide greater diversity. The experience so far is that farmers are more interested in NWFP species in order to generate income, and less interested in developing plantations focused on ecological benefits because they are not allowed to harvest this type of plantation. Most NWFP plantations are monocultures, with less ecological benefits.

Combating Desertification Program

The Combating Desertification Program (CDP) was designed as a major effort to reduce sand storms and improve the ecological environment and living conditions in seriously affected areas including Beijing, Tianjin, Hebei, Inner Mogolia and Shanxi (see Figure 5.1). The area involved is 100,120 km^2 in seventy-five counties. This program was launched in 2002 after a two-year pilot phase and will be implemented to 2010 with a planned budget of 55.9 billion *yuan*. Major strategies of the (CDP) include planting trees and grass, constructing irrigation facilities and relocating farmer and herdsman households from seriously damaged areas to allow grass and tree regeneration. If successful, it will also contribute to improving the environments of neighbouring Korea and Japan that have suffered from a series of sand storms from north China. By the end of 2005, 2.6 million ha of tree plantations were established, many farmer and herdsman households were relocated and degraded forestlands and grasslands had begun to regenerate.

Shelterbelt and Windbreak System Program

The Shelterbelt and Windbreak System Program (SWSP) consists of six subprograms. They were all launched before the mid-1990s, but some have since been expanded in terms of capital investment and area. The sub-programs are:

1. Three-North Shelterbelt (Figure 5.2)

2. Conservation Forests
    - Mid-Upper Yangtze River Watershed Conservation Forests (Figure 5.2)
    - Pearl River Watershed Conservation Forests
    - Coastline Windbreak Forests (Figure 5.2)
    - Plains Greening
    - Taihang Mountain Greening (Figure 5.2)

The Three-North Shelterbelt and Mid-Upper Yangtze River Watershed Conservation Forests subprograms are discussed below; information about other subprograms is provided in Table 5.1.

Three-North Shelterbelt

This is the earliest environment-focused forest program in China. It was launched in 1978 in 596 counties of thirteen provinces and municipalities in northeast, north and northwest China (Three-North) and involves an area of about 407 million km^2 (forty-two percent of China's land area). It is a very long-term forest program, being implemented in three stages to 2050. The first stage ended in 2000. This sub-program aims to prevent and control desertification and to restore desert lands by establishing tree plantations as shelterbelts.

By 2005, 25.1 million ha of trees had been established—18.0 million ha of tree plantations and 7.1 million ha of natural regeneration forests through mountain closure (*China Green Times* 2007b)—earning a reputation as the largest reforestation program in the world. The shelterbelts have had some environmental impacts through degraded land restoration and agri-ecological improvement in the Three-North area.

Mid-Upper Yangtze River Watershed Conservation Forests

This subprogram was initiated in 1989 in the relevant counties of eleven provinces (Table 5.1). Partly overlapping with the Natural Forest Protection Program in geographical area and program elements, this subprogram aims to protect headwaters and rehabilitate habitat in the upper and middle reaches of the Yangtze River. Its key elements include improvement in tending existing middle-aged and young natural forests and tree plantations and the establishment of new tree plantations, whereas NFPP's focus is on tree plantation establishment and a ban on logging natural forests.

Yunnan and Qinghai are among the provinces where this subprogram is implemented. This highlights its regional significance in providing eco-services such as the Desertification Combating Program. The subprogram will also benefit other Greater Mekong countries (Myanmar, Laos, Thailand,

Figure 5.2 China's Shelterbelt and Windbreak System Program

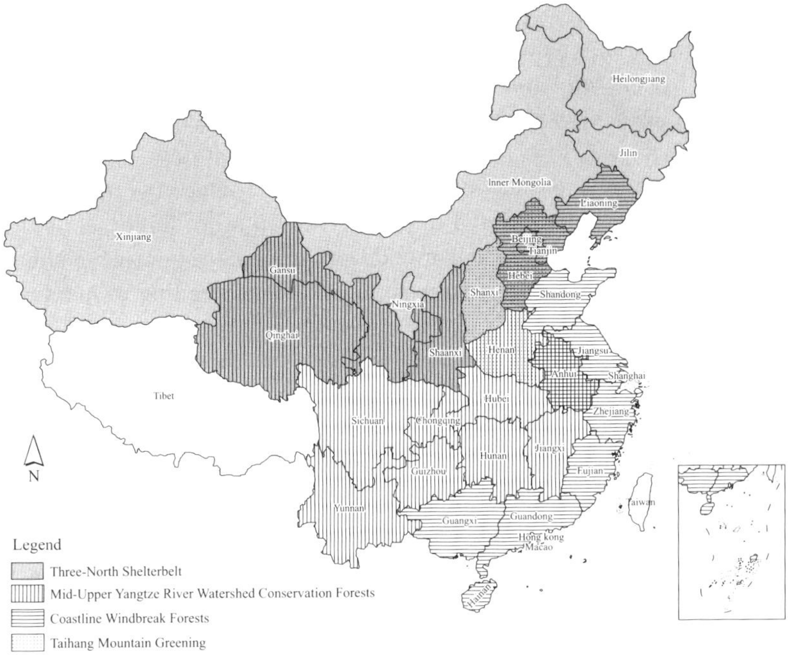

Cambodia and Vietnam), because the Mekong River originates and runs through Qinghai and Yunnan to these countries. The Thanlwin (Salween) River and Ayeyarwady River and their tributaries originate in Yunnan and Qinghai/Tibet. The Red River originates in Yunnan and runs into Vietnam.

Environmental impacts of this subprogram also include biodiversity conservation because Yunnan is a biodiversity 'hot spot' and a biological corridor of flora and fauna that runs between Yunnan and neighbouring countries.

Reforestation for wood production program

China is one of the world's leading wood-consuming countries and is ranked first in timber imports. Although wood consumption in China is relatively low per capita, with a population of 1.3 billion the overall demand for forest products is enormous. The imbalance between supply and demand for wood from Chinese forests has become greater as the Chinese economy has expanded. There has been growth in house building, expansion in the furniture industry and consumption of paper has increased. Fast-growing single species plantations have been established in recent years to satisfy some of this demand.

Wood production plantations were established in China between 1950 and 1985, but the total area under plantation was small and little attention was paid to the quality of planting and subsequent plantation maintenance, so productivity was low. In 1985 a new policy promoting 'fast-growing, high-yielding' timber production plantations was introduced by the Ministry of Forestry. The policy aimed to plant trees on more fertile soils, use higher quality nursery stock and provide improved management techniques such as site preparation, fertilizer application and weeding to maximize productivity.

The State Forestry Administration continues to develop and implement policies aimed at reducing China's dependence on imported wood fiber, paper and processed wood products by encouraging industrial plantation establishment. During its Tenth Five-Year Development Plan (2001–2005) alone, China established 357,000 ha of such industrial plantations, of which 76,100 ha is fast-growing pulpwood plantations, 110,000 ha for board and panel, 60,800 ha for large-diameter timber and 110,100 ha for others (*China Green Times* 2006a).

Take eucalypt plantations in southern China for example. They are providing new sources of income for individual farmers and collectives who have the necessary management skills and access to capital and land. Opportunities for employment have been created in some areas and these are helping secure off-farm income for some of the poorest local families and poor migrants from other provinces. Farmers are also deriving income by leasing their land to wood-based industries for plantation development. Socioeconomic impacts of eucalypt plantations are generally positive and plantation development has contributed significantly to poverty alleviation in some areas (Turnbull 2007).

The primary objective of fast-growing, short-rotation plantations is wood production, but they also have ecological benefits such as carbon sequestration to help reduce global warming. Although they have much less biodiversity than natural forest, there is little evidence that the plantations will have significant negative effects on biodiversity where they are planted on cleared land or in highly degraded areas. Water use by fast-growing plantations may be a problem in drier areas. Impacts on water supplies should be monitored and if there is a problem then silvicultural practices such as tree spacing and thinning can be modified to reduce water use and conserve local water supplies. Nutrient depletion and/or erosion of the relatively fragile soils in much of southern China are of concern but these issues too can be managed by appropriate silvicultural practices such as fertilization and minimum cultivation of planting sites. A greater risk to the sustainability of plantations is the potential for serious damage by pests and diseases to monoculture plantations that lack genetic diversity (Turnbull 2007).

International aid reforestation projects

International aid reforestation projects have also contributed to reforestation and forest rehabilitation in China. Many reforestation projects aided by multilateral organizations, foreign governments and civil organizations have been implemented in China since the 1980s.

In a series of projects beginning in 1985 the World Bank helped China address its growing need for wood products. The projects aimed to raise forest productivity and improve resource use efficiency. Between 1990 and 1997 the National Afforestation Project (NAP) was implemented to finance afforestation in sixteen provinces and over one million ha were planted. A feature of NAP was the involvement of collective forest farms, share-holding forest farms and individual households, rather than state forest farms. Complementing and extending NAP, the Forest Resource Development and Protection Project (1996–2000) aimed to expand the supply of commercial timber and pulpwood by establishing 620,000 ha of intensively managed plantations. In 2002 the China forestry program was the largest in the World Bank, and the largest forestry investment by any donor in any one country with loans, grants and credits of US$804 million (World Bank 2002). These were followed by the Poor Area Forestry Development Project and the Sustainable Forestry Development Project. These projects together helped China develop 3.2 million ha of plantations (Xinhua News Agency 2006).

Japanese loans to China started in 1980 and those for reforestation and forest rehabilitation began in 2000. The Green Obuchi Fund was implemented in 2000 and helped establish 18,800 ha of tree plantations in twenty-two provinces in China by 2005 (*China Green Times* 2005).

The joint China-Germany Reforestation Project, a ten-year project, involved an agreed investment of 2.01 billion *yuan* and contributed to an increase in forest resources. Other major bilateral aid projects in reforestation included those initiated by Australia, Canada, the Netherlands and Korea.

Environmental impacts of the reforestation programs

These reforestation programs have helped conserve existing forests and biodiversity and improve the condition of degraded lands and forests since the 1980s. A significant increase in forest and other vegetation cover and the restoration of degraded forests, including improved forest quality, is evidence of this progress. The forest area of China was 115.3 million ha in the 1977–81 NFI period and 174.9 million ha in the 1999–2003 NFI period (with some shrub lands included), an increase of forest cover from 13.0% to 18.2% (SFA 2005). This increase was achieved largely by establishing tree plantations. Tree plantations made up 31.5% of total forest area in the 1999–2003 NFI period (SFA 2005) compared to 24.7% in the 1984–88 NFI period (MoF 1989).

Though more data and research are needed to assess the extent of ecological and environmental benefits, the increase in forest and other vegetation cover has clearly had positive impacts, for example, on carbon sequestration, soil and water erosion, sand storms, soil fertility, microclimates and agro-ecological conditions.

Climate change is a major global concern, and responsive strategies include reduction in carbon emissions and tree planting. The increase in tree plantations, achieved through the forest programs, is contributing to emission cuts, as forested land is most effective in absorbing carbon dioxide and producing oxygen. It is reported (Anon. 2002) that one hectare of tropical forest fixes 4.5–16 tonnes of carbon dioxide; temperate forest, 2.7–11.25 tonnes; cold temperature forest, 1.8–9.0 tonnes; and grassland, about 1.3 tonnes. China's tree plantations are distributed in areas from subtropical to temperate zones. These extensive forests will make a considerable contribution to carbon sequestration and mitigation of global warming. This benefit will increase as current young and middle-aged forests grow and become mature.

The increase in land cover in forest and other vegetation types is beneficial to soil and water erosion control. Certainly, natural and mixed forests are able to contribute more to environment restoration and more ecological benefits than monoculture plantations by, for example, carbon sequestration, improving agri-climates and reducing runoff and soil erosion to decrease siltation of rivers and reservoirs. If planned and implemented properly, tree and grass planting is an effective response to land degradation and a means of environment restoration. Empirical evidence from the Three-North Shelterbelt program and other studies confirms that tree plantations help control soil erosion, reduce runoff and maintain and improve soil nutrients and structure if native and indigenous species are used and mixed plantations developed.[4]

The increase in land cover (trees and grasses) achieved by the Three-North Shelterbelt Program, the Farmland-to-Forest Program and the Combating Desertification Program, has begun to show its effectiveness in controlling sand-storms and reducing the damage they cause. Some cases from the programs demonstrate this ecological benefit, and that appropriate species mixtures are better than single species in developing plantations.

Evidence from the program areas indicates that the increase in forest and other vegetation cover is contributing to an improvement in microclimate and agro-ecological conditions, including soil fertility, moisture status, structure and temperature. Some 17.6 million ha of farmland benefit from the Three-North Shelterbelt Program (*China Green Times* 2007b).

The environmental impacts of China's great reforestation programs are not restricted to China but will extend to other countries. Carbon sequestration has a global impact through contributing to global warming reduction. Tree and grass plantations established in north China will reduce the impacts of sand storms on Korea and Japan. Protection and rehabilitation of forests in Yunnan, Qinghai and Tibet will benefit downstream countries including Myanmar, Laos, Thailand, Vietnam and Cambodia, through which the Mekong, Salween and Red rivers flow.

Conclusions

The three periods of extensive deforestation caused by ill-perceived policies or tenure insecurity and the over-logging from the 1950s to the 1980s resulted in destruction of forests and biodiversity and forest degradation. A significant change and evolution in the thinking about the role of forests has come about and an important shift in the management and use of natural resources has been experienced since the 1980s. Forestry policy has moved towards greater environmental protection and provision of ecological services. Massive reforestation through the formulation and implementation of reforestation forest programs following the deforestation has achieved a significant increase in forest and other vegetation cover. Forest protection programs have helped maintain existing forests and biodiversity and rehabilitate degraded forests and grasslands. As a consequence, the great reforestation activities have had, and will continue to have, many positive environmental impacts both domestically and internationally.

It is essential to maintain the current momentum of reforestation and forest protection, guided by effective policies, to maintain and expand the land cover change for greater ecological benefits. Forest policies must also consider both the country's increasing demand for wood and other forest products induced by rapid economic growth in the coming decades and the livelihoods of forest-dependent people.

Notes

1. Opinions expressed in this chapter are those of the author and do not necessarily reflect nor represent those of his organization, The Nature Conservancy. The author thanks Dr. John W. Turnbull for his inputs to the section on the Reforestation for Wood Production Program, his valuable comments on and his editing of this chapter.
2. There are two types of forestland ownership in China: state forest and collective forest. Individual trees can also be owned by individual households.
3. Although there is no reliable data on forest area and growing stock for the period before the 1970s, most forest officials, forest researchers and rural communities strongly believe that forest resources before the 1970s were much more extensive than forest resources in the 1977–81 NFI period. This suggests a significant decrease in forest area and growing stock from the 1950s to the early 1970s.
4. In general, tree plantations have fewer ecological benefits than natural forests. Some types of tree plantations can lower underground water tables and may exacerbate soil and water erosion. Single species plantations are more likely to be infected with pests and diseases. Monoculture plantations entail greater ecological risks but these can be reduced by good planning.

References

Anon. (2002), 'Roles of forest in reducing global warming', Presentation at Workshop on Reforestation/Greening and Climate Change (in Chinese), unpublished, Lin'an, Zhejiang, China.

China Green Times (Zhongguo Lüse Shibao) (2003), 'Six key forestry programs' (in Chinese), September 27: 2.

China Green Times (2005), January 17: 1.

China Green Times (2006), September 4: 2.

China Green Times (2007a), March 14: 2.

China Green Times (2007b), March 15: 1.

Clubb, E. (1972), *20th-Century China*, 2nd edition, New York: Columbia University Press.

Liu, D. (2001), 'Tenure and management of non-state forests in China since 1950: A historical review', *Environmental History*, 6 (2): 239–263.

Liu, D., and D. Edmunds (2003a), 'Devolution as a means of expanding local forest management in South China: Lessons from the past 20 years', in W. F. Hyde, B. Belcher, and J. Xu (eds) *China's Forests: Global Lessons from Market Reforms*, Washington D.C.: Resources for the Future, and Bogor: Center for International Forestry Research.

Liu, D., and D. Edmunds (2003b), 'The promises and limitations of devolution and local forest management in China', in D. Edmunds and E. Wollenberg (eds) *Local Forest Management: The Impacts of Devolution Policies*, London: Earthscan Publications.

Liu, D., O. Weng, J. Zeng, H. Zheng and Z. Zhou (2003), 'Degradation and rehabilitation of evergreen broadleaf forest in Cangwu County, Guangxi', in D. Liu (ed.) *Rehabilitation of Degraded Forests to Improve Livelihoods of Poor Farmers in South China*, Bogor: Center for International Forestry Research.

Menzies, N. K. and N. L. Peluso (1991), 'Rights of access to upland forest resources in Southwest China', *Journal of World Forest Resource Management*, 6: 1–20

Ministry of Forestry (MoF) (1989), *China Forestry Yearbook* (in Chinese), Beijing: China Forestry Press.

State Forestry Administration (SFA) (2005), *China Forest Resources Report* (in Chinese), Beijing: China Forestry Press.

Turnbull, J.W. (2007), *Development of Sustainable Forestry Plantations in China: a Review*, Impact Assessment Series 45, Canberra: Australian Centre for International Agricultural Research.

Xinhua News Agency (2006), *US$5.7 billion Foreign Funds Invested in China's Forestry Sector* (in Chinese), *http://www.gov.cn*, accessed October 21.

Yin, R. (1994), 'China's rural forestry since 1949', *Journal of World Forest Resource Management*, 7: 73–100.

Yin, R. (2003), 'Central characteristics of reforms: Measures of the effects of improved property rights, a stable policy environment, and environment protection', in W. F. Hyde, B. Belcher, and J. Xu (eds) *China's Forests: Global Lessons from Market Reforms*, Washington D.C.: Resources for the Future, and Bogor: Center for International Forestry Research.

6

Problematic Land Cover Change or Problematic Policy and Analysis?

William F. Hyde and Jintao Xu

Introduction

Our expertise is in forestry and the policies—macroeconomic and microeconomic, forestry, agricultural and food and environmental—that have induced change in forest land use since 1978. These policies, and the economic growth that has resulted from them, have been a source of major improvements in forest land use and in the critical agriculture-forestry land use margin. These improvements are often overlooked, especially by short-sighted foreign observers. However, a number of rural land use problems do remain, and it is unclear as to whether current policy is always the best means for their improvement. These perspectives will be examined in this chapter.

We begin with a review of the performance of China's forestry and agricultural sectors since 1978, the year agricultural reforms introduced China's movement toward a market economy and rapid economic growth and development. We consider the agricultural and forest land use shifts that have occurred as a result. It will become clear that, with growth and development, agricultural production has intensified, aggregate production and productivity per ha has increased and agricultural land use has declined. China has become self-sufficient in many food products. Simultaneously, its managed forests are recovering from repetitive deforestation, even as it has continued harvesting its remaining natural forests. Are there exceptions to these trends? What are the implications for the rural environment? Are there more recent policies that alter these trends or alternative government decisions that could improve on them? Are there mitigating conditions? Let us consider.

China's rural economy since 1978[1]

At the outset of economic reforms in 1978, China's forests were seriously depleted from years of severe political turmoil, economic hardship and the ill-advised attempt to produce steel in thousands of backyard furnaces. Forest stands covered approximately ninety million ha, 8.6% of China's land area or 0.10 ha per capita, among the lowest per capita shares in the world. Agricultural production was stagnant. Per capita annual income in the communes was only 62.8 *yuan* (about US$25); more than one-third of rural households were in debt, and many lived with insufficient fuel, clothing and housing. About 100 million people suffered from food shortages (Hyde, Xu, and Belcher 2003).

Table 6.1 summarizes these conditions, as well as China's growth through 1998.[2] China's real GDP increased five-fold and its per capita GDP increased almost four-fold since the introduction of the reforms. China's vast rural economy, including agricultural and forest-based communities and many other households dependent on agriculture and the forest for some of their livelihood, showed the greatest improvement. Rural per capita GDP increased a phenomenal six-fold during the same twenty-year period.

Agricultural production itself increased 225% in the first six years of reforms (till 1984) and the productivity of agricultural labor increased 172%. Agricultural land use actually declined during this period, and it has continued to decline since then. Forest sector growth has been more variable and, in some parts, not as impressive. Log production increased only slightly, and lumber production increased a moderate sixty percent. The paper industry, on the other hand, has kept pace with double-digit annual growth in the general manufacturing sector, expanding to almost five times its former level. The contrast in growth between lumber and paper is not entirely unexpected. Lumber consumption in most economies is income-inelastic, whereas paper consumption is more income-responsive and may even be elastic over a range of incomes. Therefore, we expect paper production to expand more rapidly than lumber production in order to keep up with the more rapidly increasing demand for paper that generally accompanies economic growth.

The expansion in paper production anticipates a critical partner of early economic development: deteriorating environmental conditions. The pulping technology that is an antecedent to papermaking can be a serious polluter, emitting fibrous solids and chemical waste in its effluents. In China, the pulp and paper industry has become the largest source of rural environmental pollution. Meanwhile, environmental protection has become a serious objective of the central government and, in 1999 President Jiang Zemin announced that any enterprise not in compliance with environmental regulations by 2000 would be closed.[3] More than 2000 small papermaking enterprises have already been shut down.

Table 6.1 Economic growth, 1978–1998

	1978[a]	1998	% change
GDP (1978 monetary value)			
Aggregate	362	2 312	538
GDP per capita	379	1 869	390
Rural GDP per capita	133	945	610
Forest products			
Production (thousand m^3)			
Logs	51 673	55 557	7
Lumber	11 055	17 876	62
Paper and paperboard	4 390	21 256	385
Imports (thousand m^3)			
Logs	1 870	4 820	158
Lumber	75	1 679	2 139
Paper and paperboard	767	5 760	651
Exports (thousand m^3)			
Logs	28	63[b]	125
Lumber	13	255	1 862
Paper and paperboard	229	250	9
Forest Cover			
Area (million hectares)	115	154	35
Standing volume (million m^3)	8 801	10 086	15
Agriculture			
Cropland			
Production			

Sources: China Statistics Yearbook (2000); China Forestry Development Report (2000).
Notes:
a. Imports for 1981, exports for 1983.
b. 1997 data.

The information on non-market forest products and services cannot be summarized as easily. Aggregate data on products such as fuelwood, bamboo products, mushrooms, and the many herbs collected from natural forests, and on forest-based environmental services such as erosion control, protection of biodiversity and forest recreation are simply not available. Nevertheless, the evidence that is available is instructive.

Forests are the source of perhaps forty percent of rural energy in China, but we are unsure how this estimate has changed over the course of twenty years of development. We are equally uncertain as to the effect of China's rapid growth on other non-timber and non-market forest products. We can only anticipate that the subsistence-based consumption of many non-timber forest products tends to increase with early development and then eventually decrease as development proceeds and incomes rise further yet.

Although some consumptive demands on the forests may decrease as China's economy continues to grow, we can also anticipate that nonconsumptive demands on the forest will increase. In general, as household incomes increase and overall human welfare improves, recreational uses of natural resources such as forests and demands for all varieties of environmental services also tend to increase. This too is a hypothesis based on general experience, but we are confident of its validity for China. Casual observations from any of China's mountain or lake resorts reveal a rapid increase in outdoor recreation over the past few years[4], and the central government's decisions to enforce environmental compliance and establish sustainable development as one of two fundamental strategies for the twenty-first century clearly demonstrate the increasing importance it places on environmental values.[5]

We can conclude that specific demands on the forest have been variable in their relative growth, but that the overall importance of China's forests is undoubtedly increasing. How does this expectation compare with aggregate trends in the forest inventory since 1978? Table 6.1 shows that forest cover has increased thirty-five percent, while the standing forest volume has increased a modest fifteen percent. The difference is not surprising. New seedlings and new forests grow slowly at first, and their growth in the first few years is generally too small for measurement in most forest surveys. Nevertheless, the expansion in forest cover appears to be proceeding more slowly than the increasing demands for either commercial forest products or non-consumptive forest-based environmental services.

Furthermore, net imports have not been a factor—until the logging ban in 1998, more recent than the time period of our table. The level of exports was small compared with domestic production, and the level of imports has not been much larger. Therefore, China's industry relied on substitute raw materials to absorb the difference; for example, using bamboo as a substitute for wood in some construction activities and agricultural residues as a substitute for wood fiber in paper production. Under most conditions wood and wood fiber are the preferred raw materials, and we can only anticipate that these preferences are another reason why China's forests will become even more important as the country continues to develop.

Forest organization

Let's examine the forest data itself more closely. China organizes its official forest statistics by region and administrative responsibility. Table 6.2 records the land area under forest cover for three national forest surveys conducted between 1977 and 1998 in each of six principal geographic regions and

Table 6.2 Land area in forest (forest stands in millions of hectares)[a]

Region	1977–1981	1984–1988	1994–1998	% change
Northeast				
State-owned	21.37	20.55	23.50	9.97
Collective	2.33	3.96	4.20	80.26
Total	23.70	24.51	27.70	16.88
Northwest				
State-owned	2.48	5.30	5.78	133.06
Collective	1.96	2.80	3.19	62.76
Total	7.24 (4.44)[b]	8.10	8.97	23.90 (102.03)[b]
Southeast				
State-owned	1.32	1.73	2.49	88.64
Collective	10.84	12.01	16.55	52.68
Total	12.16	13.74	19.04	56.58
Southwest				
State-owned	n.a.	8.15	9.42	
Collective	n.a.	12.24	17.39	
Total	23.53	20.39	26.81	13.94
South Central				
State-owned	1.78	2.18	2.55	43.26
Collective	10.32	14.92	22.44	117.44
Total	17.35 (12.10)[b]	17.10	24.99	44.03 (106.53)[b]
North China Plains				
State-owned	13.26	13.11	14.18	6.94
Collective	1.55	2.13	3.43	121.29
Total	14.83 (14.81)[b]	15.23	17.61	18.75 (18.91)[b]
All China				
State-owned	n.a.	51.01	62.01 (57.93)[c]	
Collective	n.a.	48.06	67.19	
Total	98.81	99.07	129.20 (125.12)[c]	30.76 (26.63)[c]

Source: National Forest Resource Statistics, various issues.

Notes:

a. China's forest survey divides 'forests' into 'economic forests', 'bamboo forests' and 'forest stands'. 'Economic forests' are tree crops like orchards and rubber. 'Forest stands' are what most of the rest of the world calls 'forests', and they are what we have reported in this table.

b. In China's second inventory (1977–1981), the data from some smaller provinces were included in the total but not in the disaggregation into state-owned and collective lands. The first entry in our 'total' row for three regions is the total reported to the national inventory. The entry in parenthesis is the sum of recorded state-owned and collective lands for the larger provinces in those three regions. The difference is the disaggregated state-owned and collective area in the smaller provinces in these regions.

c. The entries in parentheses are the sums of our regional data. The larger entry is the official national aggregate.

according to its classification as either state-owned or collective. The Northeast and the Southwest, followed by the South Central region, contain the largest forests.

The state-owned forests include approximately 3000 independent forest farms and about 135 state forest bureaus. The bureaus control their own forest farms, and most of them control their own wood-processing facilities as well. The bureaus were originally set up as logging operations for natural forests, although many of their natural stands are second growth. Many of the independent forest farms were originally set up to manage plantations. The reporting hierarchy for both institutions has changed several times. They now report to local or provincial governments for personnel and planning but both are subject to the forest management guidelines of the State Forest Administration (SFA; called the Ministry of Forestry before 1998).

Collective forests are those that were managed by local forest and agricultural collectives until 1978. Since then, individual households have gradually assumed long-term contractual responsibility for more and more of the collective lands. The collectives now account for about sixty percent of China's total forestland and individual households manage about eighty percent of that.

The land areas in state-owned and collective forests have expanded gradually, but the latter have grown more rapidly in most regions. Nevertheless, standing forest volume per hectare remains larger for the state-owned forests, even in the most recent forest inventory. We would anticipate this ordering because some state-owned forests are inaccessible and biologically mature, whereas all collective forests are accessible and, therefore, almost certain to be harvested at younger ages and in smaller volumes as they attain economic maturity.[6] The standing volumes in the state forests of most countries exceed the volumes of their non-state counterparts for the same reason.

Some discussions of China's forests modify the regional organization of Table 6.2 to more closely reflect management responsibilities. For example, the Northeast and Southwest are sometimes combined (Northeast Southwest State Forest Region, or NSSFR) as the regions containing predominantly state-owned forest and as China's major sources of industrial wood. The objective of these state forests has been to support employment in the local mills. Their standing inventory declined because the procurement prices paid by the mills were too low to support reforestation. The Southern Collective Forest Region (SCFR; including the Southeast, South Central, and some of the Southwest from Table 6.2) is the next most productive region. It, too, has suffered from procurement prices that are still too low to encourage reforestation without external financial assistance. The North-Northwest Farm Forest Region (NNFFR) covers the remainder of the country. The western part of the NNFFR is a dry area that supports few forests; the eastern part is a major agricultural area. The only significant forests in this region are government-supported shelterbelts.

Agricultural and other rural policy reforms have affected the SCFR and the NNFFR but they may have had less impact elsewhere. They have led to vigorous agroforestry activities that have significantly altered the landscape in the eastern part of the SCFR. More recent reforms of the SFA and the state forest bureaus plus a few large-scale government plantations may have had greater impacts on the NSSFR.

A general summary

What does this tell us about land use change in general? It tells us that as agricultural productivity has increased and agricultural land use has intensified and declined, forest cover has replaced some agricultural land uses. Farmers have either planted trees or allowed land to revert to trees as it was no longer needed for agricultural production.

The remaining natural forests are the upland and inland forests whose condition is a concern to many. Those who have seen these forests—or the industrial operations that draw on them—as well as those with knowledge of the environmental services provided by natural forests, recognize their condition. These are often degraded forests. We share the concern for these natural forests.

However, let us place these concerns in perspective. The data from Table 6.2 show that even the natural forest has recovered in both volume and area since 1978. In fact, the inventory of natural forest had been so seriously depleted by 1978 that there existed few economic opportunities for its further commercial exploitation. This forest could no longer provide the raw material to support employment in the wood products industry. That continues to be the case, and in recognition of this fact the government is in the process of adjusting its support for many non-viable state forest operations today.

The collective forests are also recovering, and at an even more rapid rate than their state-owned counterparts. The collective forests are not growing as rapidly as agricultural production has increased, partly because trees grow slowly and it takes time for their growth to register in official national statistics. However, these new forests replaced cropland, pastures and wasteland. Farmers do not manage their forests as intensively as they manage their agricultural lands. Therefore, we can anticipate that some of the new collective forests yield relative environmental gain. Agricultural land use has been withdrawn, largely from the poorest, steepest, rockiest agricultural lands, and the new forests are less erosive uses of those lands that also provide some new services for the surrounding environment—habitat for some wildlife, protection against some additional erosion on adjacent lands, etc.

We can further surmise that some of the increased collective forest production has replaced production from the natural forests. As such, production from the collective forests has contributed to the recovery of the natural forests and will continue to do so at least for the foreseeable future. In sum, the condition of land that was either agricultural or forested in 1978 has steadily improved since that time.

Further evidence, expected objections and contrasting concerns

Where is the evidence that these trends will continue, that agricultural demands on the land will decline, that lower quality agricultural lands will be reforested and that extractive, consumptive demands on the remaining natural forest will also decline allowing it to recover?

First, these trends are consistent with the global experience for developed countries in the twentieth century. As the countries of Western Europe and North America have developed, their demands for agricultural land have declined—despite increased aggregate agricultural production—their forests have recovered, sometimes impressively, and their rural environments have become more attractive in all respects. This, however, is the case for Western Europe and North America. Is there reason to anticipate a similar experience for China? Let us consider the evidence of some recent Chinese research.

Yaoqi Zhang, in his recent PhD dissertation and in related publications with various colleagues, examined the changes in managed and natural forest cover for China's island province of Hainan over the period from 1957 to 1994 (Y. Zhang et al. 2000; Y. Zhang 2001). His geographic focus has advantages. Because Hainan is an island, its boundaries are clear and its limited size means that only a minimum of natural and institutional variables affect any statistical observations. Zhang's most important observations for our purposes are that managed forest cover expanded in response to growth in the general economy and the population, as well as to increasing wood prices and improved land tenure. The two general economic factors (population and the aggregate economy) were responsible for twenty-five percent of the increase in managed forest cover. Of course, improvements in land tenure are institutional adjustments that also tend to occur as a product of economic growth. Furthermore, as the prices rose for agricultural crops (rubber and fruit) that compete with forests for land, these prices induced more intensive rubber and fruit production rather than extensive extraction of these products from the natural forest. Finally, at the same time, legal restrictions and other social pressures bought additional improvements in rural environmental conditions.

In sum, Zhang's observations for Hainan are entirely consistent with the Western European and North American experience. They are solid reasons to expect a similar experience in all of China as the entire country continues its economic growth. These really are not surprising results. As China's household incomes have risen, the population has moved off the farm and a larger portion of the total population now lives in towns and cities. Moreover, many of China's households have switched from wood fuel for cooking and heating as their incomes have risen. However, a visit to any of China's natural areas will reveal vast increases in outdoor recreation in recent years. The decline in consumptive uses of China's natural forests and an increase in all values placed on the rural environment is the obvious conclusion.

Nevertheless, we might anticipate the rebuttal that, for China, the improvements we observe in rural land use and the environment may be due to the enlightened activities of the central government. It is true that vast reforestation programs such as the Green Great Wall may deter some amount of wind erosion and desertification, both of them very serious land use problems for China. Other large programs, such as the proposed hydrologic control, wetland restoration, and reforestation project in Heilongjiang Province may deter future flood damage and also improve environmental amenities (for example, between 1978 and 2000 fourteen large projects of the central government accounted for an official estimate of over thirty-five million ha of newly planted trees (D. Zhang 2003)). However, the role of the central government, while still large, has been declining in China. What can we expect from small landholders as the role of the central government declines further?

Runsheng Yin (Yin and Hyde 2000; Yin 2003) provides some perspective on this question. Yin used an agricultural production function in an econometric assessment of farm household responses to the effects of the improved property rights and the liberalization of the agricultural markets that occurred under China's market reforms. He observed that, for the grain producing North Central Plains, improved property rights explained almost sixty percent of a phenomenal, more than seven-fold, increase in grain yields. New forest cover, mostly in shelterbelts and inter-cropping with agricultural crops, doubled at the same time and these household conservation investments in trees explained another five percent of the increase in agricultural production in the first six years of agricultural reforms, and twenty percent of the increase thereafter.

Yin's evidence supports the contention that China's farmers do understand the merits of long-term conservation improvement for their lands and that they have responded accordingly. As their property rights improved and as they reap the benefits of increased output, they adjust their land use practices and make the conservation improvements that increase the quality of their lands.

Finally, we must not overlook the effect of one more important result of China's reforms, an effect we mentioned briefly above. The rural nonagricultural sector has also expanded and, since about 1985, it has expanded more rapidly than the agriculture sector, growing at a rate of more than twenty percent annually between 1978 and 1998 and absorbing previously underemployed agricultural labor at an annual rate of approximately fifteen percent. Of course, the departure of underemployed agricultural labor is one more reason for the decline in agricultural use of marginal lands, and for the reversion of these lands to uses that are generally less disruptive of their environmental services (Hyde et al. 2003a).

Objections

Despite these improvements in rural land use, China lost nine million ha of its remaining natural forest between 1978 and 1998 (Rozelle et al. 2003). This is an immense number, almost eight percent of its total forest cover in 1978 and an unknown but much larger share of its remaining natural forest at that time. This represents an undeniable loss of native habitat. Beyond this fact, however, there is disagreement on the impact on wildlife habitat and biodiversity because China also experienced a net gain of eighteen million ha of forest.

Some observers are more concerned with the loss of biodiversity associated with losses of old growth forest and with the clear-felling of natural forest that may be replaced with monoculture plantations (Smil 1993; Rozelle et al. 1998; Harkness 2001). Our experience does not support these concerns. First, we have observed very few harvests of old growth and very few clearcuts in our visits to China's forests. In fact, very few old growth forests remain in China today. Most of the old growth was felled long ago. The forests we observe now are almost all secondary forests—often naturally grown, but secondary forests nonetheless. Second, selective harvests are much more common than clearcuts when the objective has been the removal of industrial timber. In fact, high-grading has been the most common practice for timber harvesting in China. High-grading removes the best trees and leaves the lowest value trees and species standing. Professional foresters detest high-grading because it leaves the worst trees as the basis for future production, but high-grading also leaves the sort of dog-eared gnarly old trees and younger trees of commercially undesirable species that create good habitat for a diversity of wildlife. Third, it is true that many of the new forests are the result of replanting with seedlings of one, two, or three species. Sometimes the new seedlings are of introduced species that were not native to the area, but usually these are planted among the remaining high-graded stand, or else they are planted on open land with no remaining standing forest. In the former case, the advantages of high-grading for biodiversity remain; in the latter case, even these forests tend to support a variety of wildlife as they mature. For example, the introduction to the World Wide Fund for Nature-China annual report for the year 2000 points to the return of bird life not seen in twenty-five years to the reforested lands within a few hours drive of Beijing (Harkness 2001).

Perhaps a better way to look at the changes in China's forest cover would be to recognize that the drawdown of natural capital like forests is a well-recognized component of economic development. Furthermore, after some level of development, most countries tend to improve their natural environments, including their forests.[7] Thus the greatest losses in natural habitat and biodiversity occur in the early stages of development, and there is a special need to identify and protect critical habitat and the most threatened categories of natural environment during this period. For China, it means that current losses in natural forest cover may be greater than they will be in the future. In some ways this would be fortunate. But it also means that the preservation of biodiversity deserves greater attention now because it is also more difficult now than it will be in the future.

Contrasting concerns

If we feel that land conversion at the margins of agricultural production, and also the pattern of forest removal are not serious concerns, then do we have other concerns for rural land use in China? Indeed, we do. Consider China's problem with desertification, its less recognized problems with degraded grazing lands, the problems bought about by increased use of agricultural pesticides, and the environmental degradation caused by the pulp and paper industry.

Desertification

Wind erosion and desertification are a long-standing problem in northwestern and north central China. Sand and desert cover 2.5 million km^2 or twenty-seven percent of China's land area, and desertification annually claims another 2000 km^2 of landscape (Gluckman 2000). The government responded in 1979 with the Three North Forest Protection Project (commonly known as the Green Great Wall), an afforestation project that now extends to 24.3 million ha. The government has added several smaller afforestation and shelterbelt projects since 1979, voluntary citizen effort has added another 5.3 million ha, and many farmers have planted their own shelterbelts. There is evidence of progress, yet desertification remains a paramount environmental problem.

Degraded grazing lands

The demands of China's domestic livestock are different. Once more, rising household incomes have increased the demand for meat. The resulting demands for forage and fodder for livestock are unknown to us. However, we can identify two important policy effects.

First, China's food security policy restricted imports of meat and replaced them with domestic production. This restriction was not important for pork, for which domestic production has always been sufficient. It was important for both beef and mutton. It meant restricting low-cost imports of these products from central Asia and feeding more domestic livestock on the marginal and environmentally risky grasslands and scrub forests of western China. The extent of the land uses losses is conjecture. There are anecdotal accounts of rodenticides being used to control pikas and other rodents over very large areas on the premise that these small animals are responsible for some of the observed overexploitation of grasslands. However, the burrowing activities of these animals almost certainly play a role in maintaining the grassland productivity. It is more likely that land degradation can be attributed to the livestock themselves and this conforms to our own observations during travel in these areas.

Second, neither the household contracts for cropland nor forests included grazing lands. The use rights for these common lands went unchanged in the 1980s and 1990s. Therefore, the common lands did not benefit from the improved household incentives and the longer-term conservation investments by individual households on other lands. The grazing lands continued to be degraded until the new Western Regional Development Program in 1998 began the transfer of some of the common pastures and scrub forests to local households, and the Natural Forest Protection Program prohibited grazing on the steepest slopes and provided incentives for their reforestation. These programs are in their infancy. It will be important to observe their effects on the common grazing lands in the years to come.

Agricultural pesticides

As we travel through the countryside, we notice considerably less birdlife in China than in many other countries we have visited. The new collective forests should support increasing amounts of wildlife. Does the demand for wildlife trade cause farmers to harvest whatever wildlife that does remain?[8] Or, as a second alternative, do China's more productive and more prosperous farmers now use more pesticides on their crops than before? The answer is essential for the long-term wellbeing of these lands.

The pulp and paper industry

We do not believe the pulp and paper industry's draw on the forest is as serious a problem as some perceive. However, the effects of the damaging effluents from paper production on the rural environment are a greater problem than many foreign observers perceive.

Pulp and paper mills are immobile, capital intensive operations. Their woody raw material represents a very small share of their total cost (and even their variable costs) of operation, and the share due to virgin wood is even smaller. Therefore, as they begin to exhaust local supplies of naturally grown wood fiber they tend to enter into agreements with local landholders to grow pulpwood. Some also initiate their own forest plantations. They do not shut down and move. We observe the same experience around the world—in Finland, in the Southern US, in South Africa, in Canada and in Indonesia. Pulp mills in the coastal plain of Virginia in the US, for example, ran out of natural forest in the early 1920s, yet they continue successful operation today, relying on farm grown virgin fiber plus wood residues from a number of sources. Indeed, the general trend for pulp mills, unlike the sawmill industry which is more mobile, is to become a market for sustainable forest production.

The environmental effluents created in the production process are a greater problem. The pulp and paper industry is the largest source of rural environmental pollution, contributing ten of China's industrial wastewater emissions and one-fourth of its chemical oxygen demand. The government has experimented with a number of mechanisms for controlling these effluents, including pollution taxes, transferable pollution rights, and closure of more than 2000 of the most polluting small mills. China's pulp and paper industry relies on non-wood fibers (generally agricultural residues) at a 3:1 ratio over wood fiber, but agricultural fibers are a greater source of both chemical and biological oxygen demand (Xu 1999). As China's economy continues to grow, its demand for paper will also grow—rapidly. This means greater pressure on the environment from the effluents of its pulp mills—and all the more pressure if international environmental preferences prevail and these mills are forced to forego their access to some wood fiber and to increase their reliance on non-wood fibers.

Protected forest reserves

China created its first national forest park, the Zhangjiajie Forest National Park, in 1982. By the end of 1999, 1118 nature reserves had been established on 86,410,000 ha (BWG 1997; SEPA 1998; FAO 2001). The expansion of the land area in nature reserves might yield benefits for biodiversity and for ecotourism and watershed management, but the limited evidence suggests that the quality of management is generally poor and highly variable (MacKinnon et al. 1996; SEPA 1998).

The finances allocated for the development of nature reserves are usually one-off investments. Adequate funding for continuing operational expenses is not generally available. Therefore, the work units responsible for each reserve are encouraged to develop their own economic ventures to generate the funds to support the operation and maintenance of the reserve. As a result, many commercial ventures have been established within the protected areas: zoos, amusement parks, breeding programs to produce endangered animals for sale, sales of mounted specimens of rare species, illegal harvest and cultivation of timber, poaching of wildlife and uncontrolled development of ecotourism. All of these ventures bring financial resources into the system and, thereby, help finance the objectives of protected area management. However, many of them also conflict with the declared objectives of protected areas (Sayer and Sun 2003).

The state forests

We have focused on China's collective forestlands. These have experienced the greatest impacts from China's reforms and the aggregate area of collective forest cover has grown substantially since 1978. However, state management still accounts for more than forty percent of China's forest and these state forests have been constrained by the managerial organization of the state-owned forest enterprises, the majority of which are in financial arrears.

The central government's most recent broad policy programs, the Western Regional Development Program and the Natural Forest Protection Program of the late 1990s, are designed, in part, to address the organizational, financial, and personnel problems of these enterprises, along with the environmental problems of the state-owned forests. These are the next large tasks for China's reforms. Success in addressing them remains problematic and speculative to this day.

Conclusion

Our review has not been comprehensive. We have focused on rural land use and, within this topic, on the margins of forest and agricultural land use. This is the area of our professional expertise and it is also the focus of both domestic and international concerns for China's rural lands.

We are relatively optimistic about the trends in these forms of land use. The land area under crops has declined since reforms began in 1978 and under managed forests have increased. The new managed forests take some pressure off the remaining natural forest and this too has increased, although at a more modest rate. These trends are consistent with general economic growth elsewhere in the world and we expect they will continue as China's economy continues to mature. They might continue at an even more rapid pace if forestry did not experience higher than normal taxation and an unusual degree of uncertainty resulting from uneven local government administration of the sector (Liu et al. 2001; Yin 2003; D. Zhang 2003). Our greater concerns are with a series of environmental abuses of the rural landscape as outlined above. These are 1) desertification; 2) degraded grazing lands; 3) excessive use of agricultural pesticides; and 4) the effluent discharges of pulp mills.

The government has programs designed to combat all but the third of these issues. It has made progress on the first but China still loses 2000 km^2 to desertification each year. Its new Western Regional Development Program has begun to address the second issue, but a modification of outdated food security policy would also help. The third and the fourth issues are more problematic. Pulp mill effluents are only indirectly a land use problem. They become worse if the pulp and paper industry has to rely more heavily on agricultural residues than wood fiber as a raw material. Will the government's new forestry programs affect this favorably, or will international environmental pressure to restrict wood imports affect it unfavorably? Each of these issues deserves our attention in the forthcoming years, and the policies that affect them deserve scrutiny.

Notes

1. This topic is explained more completely and cited more fully in Hyde, Belcher and Xu (eds) (2003).
2. 1978 marks the introduction of reforms, and it is also a year for which China has a complete inventory of forestland and standing forest volume.

1998 marks China's most recent complete forest survey. Therefore, these years comprise our best benchmark years.
3. In his address to the annual workshop on Population, Resources and the Environment on March 13.
4. The general tourism data for China show 777,710 persons visited in 1982 (China Statistical Yearbook 2002) but 83 million visits to forest parks alone in 2001 (SFA 2002).
5. Premier Zhu Rongji, in the Government Report to the National People's Congress on March 5, 1999.
6. In fact, the inventories of *accessible* state-owned forests were also depleted severely. Of 131 state-owned forest bureaus in the Northeast and Southwest in 1978, twenty-five had exhausted their mature wood resources by 1993.
7. Ruiz Perez and his associates (2003: 19) trace this pattern of development for China's bamboo forests.
8. Sayer and Chang (2003) review the demands for wildlife.

References

BWG (Biodiversity Working Group) (1997), *Conserving China's Biodiversity: Report of BWG to the China Council for International Cooperation on Environment and Development (CCICED)*, Beijing: CCICED.

China Forestry Development Report (2000), Beijing: China Forestry Press.

China Forestry Yearbook (1986), Beijing: China Forestry Press.

China Statistical Yearbook (2000; 2002), Beijing: China Statistical Press.

FAO (Food and Agriculture Organization of the United Nations) (2001), *Global Forest Resources Assessment 2000*, Forestry paper 140, Rome: FAO.

Gluckman, R. (2000), 'The desert storm', *Asiaweek*, October 13: 36–40.

Harkness, J. (2001), 'Introduction', in *WWF-China Annual Report 2000*, Beijing: Worldwide Fund for Nature.

Hyde, W. F., J. Xu and B. Belcher (2003a), 'Introduction', in W. F. Hyde, B. Belcher and J. Xu (eds), *China's Forests: Global Lessons from Market Reforms*, Washington: Resources for the Future: 1–26.

Hyde, W. F., J. Xu and B. Belcher (eds) (2003), *China's Forests: Global Lessons from Market Reforms*, Washington: Resources for the Future.

Hyde, W. and J. Liu (2001), 'NFPP and its effect on the forest sector', unpublished report completed for the Chinese Academy of Forestry (under United Nations Development Programme project CPR/00/121).

Liu, J., W. Xiao, Y. Hou, C. Xie, X. Zhang, G. Wang, X. Zhang, X. Huang and J. Li (2001), 'Study on the forestry taxation and charges system in South China Collective Forestry Region', unpublished background paper, London: International Institute for Environment and Development.

MacKinnon, J., M. Sha, C. Cheung, G. Carey (ed.), Z. Xiang, and D. Melville (1996), *A Biodiversity Review of China*, Hong Kong: World Wide Fund for Nature.

Rozelle, S., L. Guo, M. Shen, A. Hughart and J. Giles (1998), 'Poverty, networks, institutions, or education: Testing among competing hypotheses on the determinants of migration in China', Working paper, Davis: University of California, Department of Agricultural Economics.

Rozelle, S., J. Huang and V. Benzinger (2003), 'Forest exploitation and protection in reform China', in W. F. Hyde, B. Belcher and J. Xu (eds), *China's Forests: Global Lessons from Market Reforms*, Washington: Resources for the Future: 109–134.

Ruiz Pérez, M., M. Fu, J. Xie, B. Belcher and M. Zhong (1996), 'Policy change in China: The effects on the bamboo sector in Anji County', *Journal of Forest Economics*, 2 (2): 149–176.

Ruiz Pérez, M., M. Zhong, B. Belcher, C. Xie and M. Fu (1999), 'The role of bamboo plantations in rural development: the case of Anji County, Zhejiang, China', *World Development*, 27 (1): 101–104.

Ruiz Pérez, M., M. Fu, X. Yang, and B.Belcher (2001), 'Towards a more environmentally friendly bamboo forestry in China', *Journal of Forestry*, 99 (7): 14–20.

Sayer, J. A., and C. Sun (2003), 'The impacts of policy reforms on forest environments and biodiversity', in W. F. Hyde, B. Belcher and J. Xu (eds), *China's Forests: Global Lessons from Market Reforms*, Washington: Resources for the Future: 177–194.

SEPA (State Environmental Protection Agency) (1998), *China's Biodiversity: A Country Study*, Beijing: China Environmental Sciences Press.

SFA (State Forest Administration) (2002), Press conference release for First China Forest Landscape Resource Exhibition and Tianmu Mountain Forest Festival, June.

Smil, V. (1993), *China's Environmental Crisis: An Inquiry into the Limits of National Development*, New York: M.E. Sharp.

Xu, J. (1999), 'China's Paper Industry: Growth and Environmental Policy during Economic Reform', unpublished PhD thesis, Blacksburg: Virginia Polytechnic Institute and State University.

Yin, R. (2003), 'Central characteristics of reform', in W. F. Hyde, B. Belcher and J. Xu (eds), *China's Forests: Global Lessons from Market Reforms*, Washington: Resources for the Future: 59–85.

Yin, R. and W. F. Hyde (2000), 'The impact of agroforestry on agricultural productivity: the case of northern China', *Agroforestry Systems*, 50: 179–194.

Zhang, D. (2003), 'Policy reform and investment in forestry', in W. F. Hyde, B. Belcher and J. Xu (eds), *China's Forests: Global Lessons from Market Reforms*, Washington: Resources for the Future: 85–108.

Zhang, Y. (2001), 'Economic transition and forestry development', unpublished PhD thesis, Helsinki University Department of Forest Economics.

Zhang, Y., J. Uusivuori and J. Kuuluvainen (2000), 'Econometric analysis of the causes of forestland use/cover change in Hainan, China', *Canadian Journal of Forest Research*, 30: 1913–1921.

7
Collaged Landscape: History and Political Ecology of Forests in Yunnan

Abe Ken-ichi

Introduction

Contemporary China is characterized by intensive land use. Where possible, land has been thoroughly exploited to provide for its huge population. Historian Mark Elvin (Elvin and Liu 1998) has said that China's three thousand year history is a history of the destruction of its nature. The land in China has constantly been reformed throughout its history. Chinese farmers have expanded their farmlands into mountainous regions, thoroughly exploiting all available resources to boost food production.

Chinese civilization is said to be the most forest-dependent (or forest-*sacrificing*) civilization (see Bray 1984 for example). As a result of continued inroads, forest lands accounted for only 7.9% of China's total area in the 1950s, a level comparable to that of desert countries.

Even in the 1990s, forest coverage in China was only fourteen percent, less than half the global average. Today, China remains one of the countries with the poorest forest resources. Various forestry policies have been formulated and implemented over the past half century to address this problem. Nonetheless, political and social circumstances have facilitated continuing deforestation. In recent years, however, China's focus on forests has reached its keenest point in history, due to the surge in demand for timber on the one hand and the growing awareness of environmental conservation on the other. Although these trends account for only a small part of China's long history of forest use, this period marks a significant turning point for China's forests.

This chapter is comprised of two parts. Firstly, changes in forest use since the establishment of the People's Republic of China are divided into several periods. Events associated with substantial changes in vegetation are examined for this purpose. Although shifts in forestry policies are included in such changes, not all are necessarily important. The division into periods as stated in this chapter is based on changes in forest conditions and not simply on shifts in forestry policies.

Secondly, this chapter discusses policies that establish use rights, termed 'three-fixed undertakings' or *sanding qiye*,[1] especially to *ziliu shan*[2] ('self-retained mountains') as an important turning point in forest history. 'Self-retained mountains' refers to a forestry policy whereby forestlands previously operated collectively were allotted to farmers, with their management being transferred to individuals. In this chapter, special attention is paid to self-retained mountains, as this policy reflects the local people's awareness of forests and their forest management practices.

The allotment and management of designated self-retained mountains differs greatly from district to district in China's broad and diverse terrain, depending on the social, cultural and economic conditions of respective districts. This chapter focuses on the situation in Yunnan Province.

Changes in forests

Tanaka (1990) divides the development of Chinese forestry policies after the establishment of People's Republic of China into four periods: 1) from 1949 to 1957 the foundations for forestry were formed; 2) from 1958 to 1965 forestry underwent a development period; 3) from 1966 to 1976 Chinese forestry policies failed in the chaos of the Cultural Revolution; and 4) beginning in 1976 and extending to the present day forestry has been promoted.

During the formative period, foundations for forestry policies were finally established after a century of war and internal division and strife. Ownership of planted trees was granted not to the land-owner but to those who planted the trees in an attempt to encourage tree-planting by individuals. However, deforestation occurred at a greater rate than reforestation due to the expansion of farmlands, in particular through the burning of plains. The *fengshan yulin*[3] ('enclosure of mountains for forests') policy creating forest reservations on mountains was promulgated to limit cultivation as a means of recovering vegetation in neglected mountainous regions. In the development period, the government undertook full-scale reforestation projects. According to Tanaka (1990), tree-planting was promoted systematically on a large scale. Seeding by aircraft[4] was typical of large-scale reforestation projects in this period. Regarding tree-planting in urban and settled rural areas, *sibian lühua*[5] ('greening of four margins') was promoted emphasizing the greening of houses, villages, roads and riverbanks.

Eucalyptus was selected and planted extensively in Yunnan. These rows of old trees still line the roads today. In the failure period, the managerial foundations for forestry were weakened by the Cultural Revolution and then rebuilt in the promotion period in its wake. Legal systems covering forestry were consolidated when the 'three-fixed undertakings' policy was announced in 1981 and the forest law was promulgated in 1984. The 'three-fixed undertakings' policy was intended to stabilize forestland ownership'[6], establish a 'stabilized forestry production responsibility system'[7], and designate self-retained mountains. Furthermore, the large project known as 'Three Northern Windbreaks'[8] was inaugurated to combat desertification in northwest China.

Edmonds (1994) identifies three periods in forestry management in China: 1) from the establishment of new China to the beginning of the Cultural Revolution (1949 to 1966); 2) the Cultural Revolution (1966 to mid-1970s); and 3) after the Cultural Revolution (after 1978). The standards used by Edmonds in his division are unclear, yet he stresses the period of the Cultural Revolution as the 'ecological disaster' period, during which forestry policies did not function and forestlands were degraded.

This chapter takes a different approach to Edmonds (1994) and Tanaka (1990) and compiles a forest history on the basis of real changes in vegetation rather than the development of forestry policies. Social change and shifts in major policies definitely affect forests through farmers' daily life and production activities. This chapter uses results of interviews with farmers on the real major changes in vegetation in the surrounding areas of the Midu basin and when they occurred. Based on the interview results, this chapter identifies policies and social and economic changes that caused major vegetation changes.

In reconstructing the history of forests principally in Yunnan province, as explained above, two events, the Great Leap Forward and the establishment of the 'three-fixed undertakings' policy, are especially important.

The prolonged war undoubtedly caused substantial damage to mountainous lands. The Great Leap Forward dealt the finishing blow to already neglected forestlands. The small amount of forest left on the periphery of the Midu basin was uprooted at the time of the Great Leap Forward, due to the backyard steel furnaces and communal canteens typical of the movement.

The purpose of the backyard steel furnaces was to increase steel production using traditional rather than modern production methods. Backyard steel furnaces were developed in every village in China. Delivery of farmers' iron pans and pots turned into a large amount of useless scrap iron. Forests were cut down to supply fuel to this delusional alchemy.

Before the Great Leap Forward, people prepared their meals at their own homes, using readily available brushwood. Under the communal canteen system, villagers ate all three meals together. In order to prepare this considerable amount of food, large kitchen ranges and pans were provided, which required proportionately higher temperatures. The canteen workers made desperate efforts to procure the large amount of fuel resources required.

According to the statistical survey conducted in Dali (Dali Bai-autonomous County Forest Bureau 1993), Yunnan, the quantity of forest resources depleted in the twelve years from 1950 to 1961, was 7734.7 thousand m^3, while within just the three years of the Great Leap Forward, 4808.5 thousand m^3 were depleted. This survey indicates rapid and extensive deforestation in the period of the Great Leap Forward. Of the quantity of resources depleted during that period, 2735.9 thousand m^3 (56.9%) were consumed by communal canteens and 641.1 thousand m^3 were burned in backyard steel furnaces. These were the two largest consumers of forest resources.[9]

The statistical survey results provide valuable material as figures related to the loss of forest cover during the Great Leap Forward period, which are recorded together with the causes, although like all statistics from that period they are of uncertain reliability. The desertification of forestlands during the Great Leap Forward period is also recorded for all areas outside Yunnan, although no comparable statistical material is available. In recent years, there have been a number of publications describing life and society as experienced during the times of the Great Leap Forward and Cultural Revolution. These also contain graphic descriptions of forest destruction. In contrast to the above-mentioned statistics, which reveal an overwhelming volume consumed by communal canteens, these publications portray the cutting down of trees as fuel for backyard steel furnaces, as cited below.

The use of forests to fuel steel furnaces has changed pastoral landscapes completely. Good farmers became constantly occupied carrying fuel and raw materials and stoking fires. At nightfall, furnaces scattered as far as the eye could see across the wide plains, brightly illuminating the night sky (Li 1994). Only steel production was emphasized, farming being of secondary concern. Farmers ran around day and night collecting firewood. Through their ceaseless search for scrap iron and firing of backyard steel furnaces, they were completely exhausted (Li 1994). These descriptions show that trees on the fields and mountains were burnt randomly during the period of the Great Leap Forward.

The period defined by Tanaka as a development period in terms of forestry policy was indeed a time when forests were destroyed most. The 'ecological disaster' stressed by Edmonds occurred most radically not in the Cultural Revolution period but during the Great Leap Forward. The Cultural Revolution period was a time when no effective measure was taken for the regeneration of forests damaged during the Great Leap Forward period. Among other things, the political confusion affected forests most, although it is certain that many natural disasters also occurred at that time.

The 'three-fixed undertakings' was the most powerful government forestry policy implemented for forest regeneration in the wake of the deforestation suffered during the Great Leap Forward and subsequent Cultural Revolution. The policy was especially effective around the basin. The allotment of some forestlands to individual farmers as self-retained mountains is comparable to the production responsibility system instituted in agriculture in 1982. Farmers were allowed to plant trees and sell their products by themselves, under the guidance and with the assistance of the government (through tree-planting loans, seedlings supplied at low cost, etc.).

As productivity increased markedly in agriculture after the allotment of self-retained mountains, it was expected that farmers, anticipating product sales, would actively plant trees, thereby bringing about increased afforestation.

Forest cover change: Yunnan

Yunnan province has varied topographic features and diversified natural vegetation. The population consists of Han and twenty other coexisting minorities, differing in history and culture. Originally Yunnan was a botanical and ethnic mosaic. In addition to this, social and economic factors have contributed to regional differences in the development of tree planting on self-retained mountains.

The allotment of self-retained mountains to farmers did not immediately lead to tree planting. In the short term, effects contrary to expectation occurred in terms of forest conservation and an increase in forested areas. Farmers felled and sold trees planted on allotted self-retained mountains as soon as they gained rights to them, as a hedge against another sudden policy shift removing their use rights.

An example of such cases was seen at the periphery of the Teng Chong basin in the southwest part of Yunnan. A plantation forest under village control was allotted as part of the self-retained mountains program in the aforementioned area, where the Han people predominantly lived. It was a mature forest of well-kept hoop pine (*Araucaria cunninghamii*) of high economic value as building material. The trees in the forest were cut down and sold by farmers immediately after the allotment in 1983, reasoning that it was better to convert the wood into money soon after gaining ownership before the government made a policy shift.

The deforestation boom was relatively common where established plantation forests had been allotted. Farmers finally began planting trees when the dust had settled. To return to the case of Teng Chong, the farmers began planting trees on their self-retained mountains in 1984, using supplies of seedlings provided by the local government at a low cost. They planted hoop pine (*Araucaria cunninghamii*), armand pine (*Pinus armandii*), and Yunnan pine (*Pinus yunnanensis*). Armand pine and Yunnan pine are bought on the local market, and it was expected that hoop pine would be sold on the international market (from interview with officials at the Tenjong County Forest Bureau).

The following is an example of the allotment of bare plots of land instead of plantation forest (Abe 1997). The Midu basin is located roughly at the center of Yunnan. Han people immigrated to the area from ancient times. The slopes of the mountains surrounding the basin were highly weathered bare lands and were allotted to farmers as self-retained mountains four years after the central government made the decision.

The land was allotted extremely roughly. Although family size and slope gradient were taken into account, the slope base was divided easily to define lots, extending borders along planes perpendicular to the slope; in other words, strips of slope were granted to families on a one-for-one basis.

The allotment was imprecise because no one thought self-retained mountains was a valuable endeavor. The farmers had no definite idea about the future of the project, or what they would gain from it. Some families preferred not to receive their allotment, being concerned about obligations and tax arising from the allotment.

Indeed, the local government made it obligatory for farmers to plant eucalyptus trees (*Eucalyptus globurus*) on their self-retained mountains. This led to the relinquishment of allocations by some farmers due to the heavy burden of tree planting involved.

The local government twice gave instructions on the planting of eucalyptus trees. The first instruction was specified immediately after the allotment of self-retained mountains (1985 to 1986). In this case, the government made interest-free loans of twenty-four *yuan* per *mu* available for purchasing seedlings. The second order was given in 1989 and 1990, when the government supplied seedlings for free. In the beginning the farmers did not actively plant trees, despite the guidance and support of the government.

The eucalyptus trees on the slope are relatively well maintained at present. Some farmers even carried out supplementary planting as appropriate after the government-led concerted planting effort, becoming positively engaged in the planting of eucalyptus trees when they found that the trees were profitable (Table 7.1).

Oil is extracted from the leaves of eucalypts. As demand for eucalyptus oil increased rapidly in the 1990s, eucalyptus leaves, the raw material for the oil, began to fetch a good price. Distillation of eucalyptus leaves became common in the basin during the winter, the agricultural off-season.

The government's intent in planting was to protect soil on slopes from runoff. The preservation of soil was not a positive motivator for farmers to plant eucalyptus trees. Farmers remained disinterested in self-retained mountains until they found they could actually earn money from the project. Cash earnings gained from the sales of eucalyptus leaves have now become very large. At present, growing eucalyptus trees for the leaves is comparable to raising pigs, which was almost the sole source of income previously.

Farmers plant trees on self-retained mountains if tree planting brings about economic benefits, as seen in planting hoop pine for building material and eucalyptus for the sale of leaves. However, direct and short-term economic benefits are not the sole incentive involved—farmers in some cases utilize their self-retained mountains for long-term goals.

A village of the Hani people is located on a plateau at a height of about 2000 m above sea level in the Ailou mountain range at the center of Yunnan province. Terraced paddy fields spread out below the village. Slash-and-burn agriculture was once practiced above the village, and formerly burnt fields are used as secondary forest areas today, although part of this is still used as fields. There is also a sacred forest zone that remains untouched.

In the village surveyed, 0.5 *mu* was uniformly allotted to each family as self-retained mountains. The remaining 200 *mu* was not divided among families, including the sacred forest area. Instead, the remaining area was placed under village control as a communally-owned forest.

The village is remote from the arterial road. It is thus not a simple task for villagers to carry timber to the market. The trees they plant on their self-retained mountains are fruit trees for personal consumption and alder trees (*Alunus nepalensis*) for fuelwood, rather than tree species of high market value. Hoop pine of high commercial value is planted in some cases, but is intended for rebuilding local houses rather than for sale. Villagers do not see their self-retained mountains as a source of cash income. Some say that their self-retained mountains are valuable as they will provide their

Table 7.1 Economic comparison between eucalyptus cultivation and pig rearing at a village in Midu Basin

No	Area of *Ziliushan* (*mu*)	Year Planted	Total Income from Eucalyptus leaves(*Yuan*) Year 1995	Year 1994	Total Income from Pig Rearing(*Yuan*) Year 1995	Year 1994	Comments by Farmer *Note
1	6	1992	2,000	1,000	3,000	3,000	PR is more profitable, but more labor intensive & more costly in tearms of fodder
2	1	1989	200	180	0	1,500	PR is more important as a source of cash income
3	3	1986 1989	1,000	700 ~ 800	1,000	0*	Total cash income from PR is same as that from EL. But PR is more labor intensive
4	0*						*decline to get *Ziliushan*
5	0.8	1982 1989	170	150 ~ 160	1,500	500 ~ 600	
6	1 ~ 2	1986 1989	200	100	-	-	
7	0*	1989	80	70	2,000	2,000	*return once allotted *Zilushan*
8	1 ~ 2	1982 1990	200		0	600 ~ 700	PR & EL are both impartant. PR is laborious but manure is valuable
9	2 ~ 3*	1987	800	450	-	-	PR is too labor intensive *Peach trees are intercropped
10	6 ~ 7	1989	300	150	-	-	
11	4+3*	1986 ~ 87	*:*	**			*Homestead garden planted with Eucalyptus tree. **Oil maker, thus EL are home consumed
12	0.4 ~ 0.5	1986 ~ 87	200 ~ 300	200 ~ 300	0*	300	EL is more profitable *due to disease
13	4+3*	1986	300	400	3,000	2,000	PR is more profitable but more labor intensive *Leased from neighbor
14	2+3*	1987	300	350	2,000	1,250	PR is more important. Price of EL is unstable *Leased from neighbor
15	1	1987	50	20	1,000	0	PR is more important
16	2+3*	1987 ~ 88	100	200	0	400 ~ 500	PR is more important * River bank. Leased from village.
17	1+0.2	1987 ~ 88	250	200	1,800	2,000	EL & PR are equally important
18	5+2.5* +0.5**	1987 ~ 88	350	350	0	1,200	EL & PR are equally important *Leased from neighbor **Field converted to Eucalyptus Plantation
19	2+0.4*	1986	200	100	0	1,000	PR is more profitable unless pigs are unhealthy * Homestead garden planted with Eucalyptus trees.
20	1+0.3*	1987 ~ 1986	300	280	750		ER & PR are equally important. EL requires less labor * Homestead garden planted with Eucalyptus trees.
21	2+0.3*	1987 ~ 88	200	110	800	1,200	*Homestead garden planted with Eucalyptus trees.
22	3+1*	1987 ~ 1985	500	620	3,800	3,500	*Homestead garden planted with Eucalyptus trees.
23	1+1*	1987 ~ 89 1985	150	150	8,000	4,000	EL require less labor but too cheap. Planning to convert into orchard. *Homestead garden planted with Eucalyptus trees.
24	2.8+0.6*	1987 ~ 88	580	700	-	-	*Homestead garden planted with Eucalyptus trees.

PR = Pig Rearing EL = Eucalyptus Leaves

grandchildren with timber for their burial caskets. To earn cash, villagers work away from home.

Many minority groups set up part or most of their allocated forest areas as sacred forests and prohibit use, as seen in the case of the Hani village. In some of these areas, people have not carried out allotment of forest as self-retained mountains or have carried it out but only as a matter of form.

A village of Thai people located on low-lying land in the southern part of Yunnan grows paddy rice as their main livelihood. Although 600 to 700 *mu* of forest was available for allotment as self-retained mountains, villagers have chosen not to divide it up for the individuals as self-retained mountains. They maintain a community with strong social coherence through the adherence to strict rules, evident in this typical governing of the sacred forests. Usage of the forest is left in the hands of ruling elders. Villagers are allowed to use it only ten days per year, in February or March in the agricultural off-season. They collect fuel wood in that period, decided upon by the ruling elders. As a result, the forest remains almost pristine.

Tibetan people living in the Zhong Xun basin on the northwest part of a plateau 3200 m above sea level grow naked barley and farm yak and other livestock as their main livelihood. Forests in the surrounding mountains were cut down during the Cultural Revolution. At present almost no forest is left, aside from the scrubland of the grassy plains. The only exception is the sacred forest called '*Shidda*'. Forestlands, including *Shidda* in the Tibetan village, have also never been allotted to individuals as self-retained mountains, regardless of vegetation such as trees, bush or grass.

Important forest resources for Tibetan people are sticks and leaves used to spread in livestock stables, as well as fuel wood. They prune and collect sticks and leaves of trees of the family *Fagaceae* in the fall, before they stable yak and cattle over the winter. Sticks and leaves mixed with livestock manure during the winter turns into valuable fertilizer. The Jondeng basin has plenty of fagaceous bush for making barnyard manure and fuel wood. On the one hand, the richness of available resources ensures self-sufficiency and on the other, no tree species are suited to the cold climate of the plateau for cultivating forestlands for profit. Thus, allotment of forestlands to individual farmers as self-retained mountains would bring them no gain.

Forestland alchemy

Regional variations in the allotment of officially managed forestlands to individuals as self-retained mountains have been examined, taking Yunnan as an example. Yunnan cannot be seen as representative, however, because of the wide variation between regions, including those where forests have not been allotted to individuals at all.

In China, forest resources were traditionally viewed as free goods of no economic merit. They had no economic value for local inhabitants, but needless to say, did have considerable *in situ* utility value as a source for collecting fuel wood, for hunting animals and for collecting mushrooms and medicinal herbs. With the shift to individual use rights, involving a number of stages, forests have come to have exchange value as a commodity.

Take Yunnan again as an example, focusing on the upstream region of the Salween River at the west end of the province. Ethnic minorities such as the Lisu and Nu dwell in the upper part of steep slopes of the deep valley formed by the Salween River. Their cultivated lands are small, steep and located at high altitudes. The area above 2400 m above sea level, or above the margin of cultivation, accounts for 69.7% of the total area of their lands. The area of sloping ground at a gradient of twenty-five degrees or more, accounts for 76.6%. Cultivated acreage and agricultural production per household is extremely low, and local farmers have few paddy fields. Maize and buckwheat are grown. Being the poorest region in China, in 1994 there were no farming households with an annual income in excess of 300 *yuan* among the 2388 households in the Shan area, where one of the villages surveyed was located.

In the mid-1990s, the farmers in the area began logging timber blocks, travelling for half a day or more into the forests. Only a few large trees remain. Cutting the timber into blocks to fit the size of their backpacks, they would carry the timber back to the village. Timber blocks would then be carried to a weekly market to sell. One block is traded at seventeen to twenty *yuan*, depending on the tree species. The blocks would then be processed into flooring materials by riverside timber mills and shipped to cities outside the province. In one of the villages surveyed, annual average cash earnings per household from the sale of timber blocks reached 171 *yuan* in 1997 and 319 *yuan* in 1998. This implies that this timber product became a valuable source of income for farming households, the average income of which was below 300 *yuan*. Forests remote from village areas previously of no economic merit for local people became profitable overnight, through a shift from utility value to exchange value, a phenomenon comparable to alchemy.

As stated earlier, demand for forest resources has rapidly increased in China as a result of high growth brought about by economic liberalization. According to Harkness (1998), for example, demand for timber increased by 196 million m^3 annually from 1973 to 1976, which further increased by 344 million m^3 per year from 1982 to 1988.

The growth in demand for timber can be a direct incentive to plant trees, as described above. Farmers plant trees once they recognize the presence of markets and possibilities of shipping timber there. Tree planting is now clearly a profitable enterprise, and China has become the world's largest tree planting country in terms of acreage.

The demand for timber has not only undergone quantitative changes, but also qualitative changes, or diversification. In urban areas in China, in particular, old buildings were pulled down and modern buildings constructed in the 1990s. Unlike conventional, rugged, utility-oriented ones, new buildings have elaborate interior decorations made possible by, for example, the abundant use of above-mentioned flooring materials.

Tree planting on self-retained mountains is expected to continue to increase in the future. It is likely that demand for timber will continue to grow for the time being. Demand for timber can be met by tree planting if quantity is the issue. However, tree planting alone is not sufficient for meeting diversified demand. Under the shade of expanding plantation forests, natural forests have rapidly decreased.

The risk of the alchemy practiced on forestlands becomes obvious if this action is considered from an environmental point of view, especially from the perspective of preservation of biological diversity. Alchemy does not produce the gold it promises. When reexamined in terms of the environment, the earth of Yunnan is turning into a homogenous or monotonous one, although at a glance it appears to be a collage of various types of forest.

Charting the future of China's forests is no doubt a national issue in terms of nationwide land conservation and of regulating the supply of and demand for timber. At the same time, such a national issue can be addressed in a feasible manner only when each district seeks optimal solutions to meet local problems. The complete picture of China's vast forest issue arguably becomes visible only when the realities of each district are understood.

Notes

1. 三定事业
2. 自留山
3. 封山育林
4. 飞播
5. 四旁绿化
6. 定山权林权
7. 定林业生产責任制

8. 三北防护林
9. After the Great Leap Forward, the principal cause of the loss of forest resources was the need for fuel for brick and roof-tile kilns. There are estimates that these cottage industries consumed approximately two million m^3, accounting for sixty-five percent of 3074.5 thousand m^3, the total forest resources consumed in 1980.

References

Abe, K. (1997), 'Forest history in Yunnan, China (II): Privately owned forests and eucalypt plantations in Han-dominated basins' (in Japanese), *Journal of Southeast Asia Studies*, 35 (30): 445–464

Bray, F. (1984), *Needham's Science & Civilisation in China: Agriculture*, Cambridge: Cambridge University Press.

Chang, J. (1991), *Wild Swans: Three Daughters of China*, London: Simon & Schuster.

Dali Bai-autonomous County Forest Bureau (ed.) (1993), *Forest History of Dali Bai-autonomous County* (in Chinese), Kunming: Yunnan Ethnic Publication.

Edmonds, R. (1994), *Patterns of China's Lost Harmony*, London: Routledge.

Elvin, M. and T. Liu (1998), *Sediments of Time: Environment and Society in Chinese History*, Cambridge: Cambridge University Press.

Harkness, J. (1998), 'Recent trends in forestry and conservation of biodiversity in China', *The China Quarterly*: 911–934.

Li, Z. (1994), *The Private Life of Chairman Mao* (in Japanese), translated by Shinjo Tetsu, Tokyo: Bungei Shunjyu.

Shapiro, J. (2001), *Mao's War against Nature: Politics and the Environment in Revolutionary China*, Cambridge: Cambridge University Press.

Tanaka, S. (1990), 'Chugoku no rinsei' (Forest policy in China), in R. Handa (ed.), *Rinseigaku* (in Japanese), Tokyo: Buneido: 252–255.

8
Rubber Planting and Eco-Environmental/Socio-cultural Transition in Xishuangbanna

Yin Shaoting

The introduction and development of rubber in Xishuangbanna

The planting of rubber trees (*Hevea brasiliensis*) in Xishuangbanna dates back to the 1940s. Then, a Thai-Chinese named Qian Fangzhou brought 500,000 rubber seeds into Ganlanba, Xishuangbanna together with two workers, but his efforts ended in nothing as none of the seeds germinated. Undeterred, in 1948 Qian carried 20,000 rubber saplings on horseback to Ganlanba. He cultivated 300 acres of rubber in Jingsongka, naming this first rubber farm in Xishuangbanna 'Xianhua Rubber Farm'. However, as a result of the unstable political situation, wars, natural disasters and destruction caused by both people and livestock in the early years, only ninety-one rubber trees in the whole farm survived (Li 2001).

As the People's Republic of China came into being in 1949, the government began to build rubber plantation farms in Xishuangbanna and Hainan to meet the needs of military and civilian industries. During the last fifty years, significant rubber plantation development has occurred in Xishuangbanna. Below is a discussion of rubber planting data in this area.

Proportional increase of rubber planting in Dongfeng Farm over fifty years

Founded in the 1950s, Dongfeng Farm has a history spanning about fifty years. The farm is located in Menglongba of Xiaojie Township, Damenglong Town, Jinghong City, Xishuangbanna Prefecture, an area that was predominately old-growth forest fifty years ago. In 1958 the farm began to fell trees and planted eighty *mu* of rubber trees (one *mu* is equal to $650m^2$), cultivating a further 127 *mu* in its nursery. The plantation increased to 17,126 *mu* in 1962, 39,135 *mu* in 1968, 74,140 *mu* in 1975, 123,348.6 *mu* in 1981 and 144,579 *mu* in 1988. As of 1999, the entire plantation area of the farm was 251,433.6 *mu*, of which rubber occupied 174,521.85 *mu* (S. Zhang 2001).

Proportional increase of rubber planting in Mengpeng Farm over fifty years

Mengpeng Farm was established in Mengpeng Town in the Southwest region of Mengla County in 1959. The farm felled enough forest to plant 1874 *mu* of rubber trees in 1959, then 5971 *mu* in 1970, 21,000 *mu* in 1974, 119,000 *mu* in 1982 and 194,156 *mu* in 1997, making it the largest rubber farm in China (N. Zhang 2002).

Proportional increase of rubber planting in Mengla County over fifty years

Mengla County lies in the south of Xishuangbanna, covering 7093 km^2. The county established a state-owned rubber plantation in 1959. There were 54,174 *mu* of rubber trees in 1969, 311,961 *mu* in 1988, and 524,096 *mu* in 1991. The peasants of this county planted 85.3 *mu* of rubber trees with the help of the state-owned farm in 1974, growing to 479,384 *mu* by 1999. That year the total area under rubber plantation in this county was 1,003,470 *mu*, comprising 9.3% of the whole county (Yang 2002).

Proportional increase of rubber planting in Xishuangbanna Prefecture over fifty years

The year 1956 witnessed the emergence of two farms: Jinghong and Guanglong, where 351 *mu* of rubber trees were planted. Ten years later there were already 120,000 *mu*, then 302,700 *mu* in 1976 and 1,000,000 *mu* by 2000.

The government has encouraged and supported local ethnic peasants to plant rubber trees while developing state-owned farms. However, according to statistics in 1970, there were only 471 *mu* of rubber trees belonging to those peasants. After that, as a result of the obvious economic profitability of the industry, they began to increase their plantings by 2,000 *mu* per year. There were already 100,000 *mu* of rubber trees in 1983, then 1,100,000 *mu* in 2002—100,000 *mu* more than state-owned farms (Li 2002).

Rubber immigrants in Xishuangbanna

Population growth is a key factor in changes in the natural environment. The relationship between rubber planting and eco-environmental transition in Xishuangbanna reflects the increase of cultivated land area as well as the immigration tide of workers involved in rubber planting. The following statistics from several districts, counties and the prefecture itself illustrate the situation of rubber immigrants over a fifty-year period.

Table 8.1 Increase of rubber planting by the state and peasants in Xishuangbanna

	Year	Planting Area (Mu)
State Farms	1956	351
	1965	120 000
	1976	302 700
	2002	1 000 000
Private Farms	1970	47.1
	1983	100 000
	2002	1 100 000
Total	2002	2 100 000

Rubber immigrants at Menglongba District over fifty years

The Agricultural Cultivation Bureau decided to build Damenglong Farm at Xiaojie Township of Menglongba District, Jinghong City in 1957, sending fourteen workers there. Several farms, including Dongfeng, were established in 1958 with the labor of 1862 immigrants from Simao, Kunming and the army. By 1959 there were already 2186 rubber immigrants in the area. There were 3129 immigrants from Liling of Hunan Province, making the Menglongba District immigrant population as large as 8958. Between 1969 and 1971, 13,522 school leavers came to work on Dongfeng Farm from Beijing, Shanghai, Chongqing and Kunming. In 1977 the population at Dongfeng Farm was 24,364. Some school leavers left the farm for their hometowns in 1979, and the population fell to 17,328 in 1980. From that time on there was little change in the farm's population.

Rubber immigrants in Mengla County over fifty years

The first rubber farm in Mengla County was established in 1958, when the population of that county was 45,628. This farm was called Manche Farm, and was initially settled by about 100 immigrants. The main farm of the county, called Mengla Farm, was established in 1963, after which the population of the county increased sharply as immigrants came from Hunan to work on the farms. In the following three years, the population increased to 2,000. In 1970, 224 immigrants from Mojiang came into Mengla, while at the same time young people from Shanghai, Beijing, Sichuan and Kunming gathered on the farms as well. This was the peak period for population increase in Mengla—the average being 4000 people per year. In 1980, the population of Mengla increased to 129,415, a net increase of 84,627. The population density in 1949 was 6.23 persons per square kilometer, but by 1980 this figure had

Table 8.2 Rubber immigrants in Menglongba District over fifty years

Year	Immigrants
1957	14
1958	1 862
1959	2 186
1961	5 396
1967	8 958
1977	24 369
1979	13 588
1980	17 328

Table 8.3 Population growth in Mengla County over fifty years

Year	Population	Density (person/km^2)
1949	44 743	6.23
1957	52 777	
1958	45 628	
1961	57 485	
1978	142 833	
1979	125 570	
1980	1 129 415	18.87
1999	194 630	27.46

reached 18.87. The population of Han also increased. The total number was 8654 in 1959, but reached 59,071 by 1975. In more than ten years, from 1980 to 1990, the total population of the whole county increased by 65,215. During this period population density increased from 18.87 persons per square kilometer to 27.46 persons (Yang, 2002).

Rubber immigrants in Xishuangbanna Prefecture over fifty years

The earliest rubber immigrants to Xishuangbanna included veterans, 'Young Volunteers for Opening up Virgin Soil' from Kunming, government officials at various levels and residents in the 1950s. By the end of 1959 there were already forty-one rubber farms with 3000 workers in Yunnan. There were 36,695 Hunan immigrants moving to Yunnan in 1960, 22,236 of whom went to Xishuangbanna. Thousands of school leavers went to Yunnan in 1965, 1606 of whom were from Chongqing and settled down at the farms in Xishuangbanna. There were 57,099 school leavers from Shanghai, 9394 from Beijing and 41,068 from Sichuan being sent to Yunnan from 1968 to 1971, a number of whom went to Xishuangbanna. Then, from 1970 to 1977,

7625 people of 1924 households in Simao moved to Xishuangbanna. The 1982 census in showed that the population of Xishuangbanna had increased by 440,300 over a thirty-year period, from 226,100 in 1953 to 646,400. The population of Han Chinese also increased from 1700 in 1955 when the Prefecture was founded to 191,700 (Cang 2002).

The socio-cultural and eco-environmental effects of rubber planting

There has been great economic profit from rubber planting in large areas in Xishuangbanna over the past fifty years, not only benefiting the nation and local government, but also the ethnic groups residing there. This is why the plantations were enlarged year after year. However, from the perspective of the eco-environment, there have also been serious consequences of the growth of plantations. Socio-cultural changes are also evident.

The most serious effect of rubber planting on the eco-environment is the disappearance of old-growth forests. According to those workers moving to Xishuangbanna in the 1950s, there were 'thick and dark forests' everywhere. Some workers went to neighboring villages, getting lost in the forests on the way. However, today there is no sign of old-growth forest in the area (Li 2002). There were 858,000 *mu* spread over the four nature reserves of Menglong, Mengyang, Menglun and Mengla in Xishuangbanna in 1958. Now after so many years of cultivation, all of the four reserves have been damaged and there are only 102 *mu* of forestland left in Menglong, making reservation an empty word (Li 2002). The total area of Xiahuangbanna is 19,700 km^2, sixty-five percent of which was covered with mainly old-growth forests in the 1950s. Then in the 1970s the rate of coverage fell to twenty-four percent. According to the statistics, in 2000 the rate had increased to forty-five percent, but this figure includes rubber trees and replanted forest, with only one third old-growth forests. The area of old-growth forestlands has declined from 11,000 km^2 to 3000 km^2 within fifty years, a loss of seventy percent. This is really an astonishing figure!

Old-growth forests, especially tropical forests, are treasure houses of biodiversity. There used to be hundreds of species of animals and thousands of species of plants in Xishuangbanna. Rubber cultivation in those forests caused the dying out of plants and animals and the destruction of the ecosystem, a most serious consequence. Old-growth forests can also act as natural 'green reservoirs', moderating the climate and mitigating natural disasters like flood and drought, producing great benefits to local people whether living in the basin or mountain areas. However, the climate in Xishuangbanna has been changing in recent years. There are no foggy days anymore, while in the past there were more than 160 days each year in fog. The change to rivers is also obvious. The half-year-long drought season reduces water levels incrementally and at times rivers are left parched, while in the rainy season floods occur. The loss of water and soil is already a significant problem in Xishuangbanna. Five-thousand square kilometers, 26.3% of the total area of the Prefecture, is facing this problem, which is mainly an aftereffect of rubber plantations taking the place of old-growth forests. Furthermore, the production process of dry rubber is usually carried out right after the harvest. Without relevant equipment for dealing with wastewater, tons of poisonous effluents from the production process are being poured into surrounding rivers and lakes, causing serious pollution problems, having a huge impact on the environment as well as people and livestock.

Ethnic groups in Xishuangbanna have experienced great socio-cultural changes over the past fifty years, due in large part to the introduction and development of rubber plantations. Before the 1950s, Xishuangbanna was a relatively closed world. After the foundation of the People's Republic of China in 1949, the government decided to develop rubber planting in this area. Immigrants from the interior built military farms in Jinghong, Menghai and Mengla. Such farms are a kind of product of industrial society, bringing in Han Chinese values and culture, highways and automobiles, modern communication and media, hospitals and schools, production methods and ways of life under a military system, administration and technology of intensive plant production industry, the market economy and the concept of consumer goods and so on. All of these changes appeared suddenly in a closed society that was still at a 'feudal serf' stage, and inevitably local people came into contact with these things. We can imagine what an assault this would be!

Planting rubber trees and building farms has had a significant effect on local people, among which the most direct influence is on the occupation and utilization of resources. In the 1950s, the distribution of ethnic groups in Xishuangbanna was in a three-dimensional form. Dai people were living in the basins and on river banks at a low altitude while Bulang, Hani, Jinuo, Yao, Mao, Yi and Lahu took the semi-mountain and mountainous areas. Rubber is a kind of tropical plant that can be cultivated in lowland below 800 m in altitude in Xishuangbanna, so all the farms were built in basin areas and the land cultivated by those farms was farmland and forests of Dai and lower slopes of the mountain ethnic groups. The scale of the developments, the influx of a large immigrant population and the extensive areas of cultivated land changed the former situation of a sparsely populated terrain, putting great pressure on the relationship between humans and land. At present, due to the lack of land and forest resources, the traditional Dai way of well-organized organic rice cultivation has had to convert to high-yield, cost intensive chemical rice cultivation. Mountain residents have also had to give up their traditional swidden cultivation, a method of agriculture with eco-preservation functions, and have adopted fixed dry land cultivation methods resulting in weed disasters and soil degradation. What's more, hunting and gathering used to be an important part of Dai and other mountain ethnic groups' livelihoods, diversifying their food and nutrition. Nowadays with severe deforestation, hunting and gathering can only exist in memory, bringing a halt to their traditional knowledge and reducing the nutritious elements in their food.

Rubber planting has also had a big influence on the transition of local ethnic groups' traditional values. In the 1950s, all the mountain residents believed in animism and revered Nature. There was a sacred mountain, sacred woods and a sacred tree and rituals for the mountain god, tree god and river god in each village. Dai people in the basin area also believed in spirits like a dragon mountain, dragon tree and practiced rituals for them. Their belief in gods has changed significantly since they began to plant rubber trees in the 1970s. Rubber planting brought them money and as they became richer, the market economy became more important. Many villages cut down their sacred woods on their holy mountain for planting rubber trees. Nowadays, rubber has become the most important economic resource in many villages. Rubber trees have become a family's main property and key heritage, even a symbol of wealth and social status. In the past girls considered farmland and house as well as love when choosing a husband, now rubber trees also play a large part. There used to be a high divorce rate in Dai society but it is much lower now, mainly because of rubber trees. According to local rules, a woman can only take the rubber trees that were part of dowry once she left her husband, while those planted during their marriage remain with the husband, which makes divorce a not so casual event any more (Du 2002).

Above all, the impact of rubber planting on various aspects of society in Xishuangbanna has been huge. In recent years, considering the ecological side-effects, Yunnan Province and the Prefectural government have taken the advice of experts, no longer allowing peasants and farms to enlarge their plantations. However, as it is already a fact that a large area of land in Xishuangbanna is under rubber plantation, it will continue to have a permanent and profound eco-environmental and socio-cultural impact.

References

Cang, M. (2002), 'Rubber immigrants on the Yunnan frontier', in S. Yin Shaoting and F. Yoko (eds), *Rainforest and Rubber Forest*, Kunming: Yunnan Education Publishing House: 7–14.

Du, J. (2002), 'Reducing the selectivity and strengthening the risk', in S. Yin Shaoting and F. Yoko (eds), *Rainforest and Rubber Forest*, Kunming: Yunnan Education Publishing House: 78–98.

He, Y. (2002), 'A brief introduction on the changes of land system and relationship of man and land', in S. Yin Shaoting and F. Yoko (eds), *Rainforest and Rubber Forest*, Kunming: Yunnan Education Publishing House: 220–243.

Li, X. (2002), 'Green giant', in S. Yin Shaoting and F. Yoko (eds), *Rainforest and Rubber Forest*, Kunming: Yunnan Education Publishing House: 260–302.

Yang, X. (2002), 'Resource utilization in Mengla: Development, reasons and prospects', in S. Yin Shaoting and F. Yoko (eds), *Rainforest and Rubber Forest*, Kunming: Yunnan Education Publishing House: 176–219.

Zhang, N. (2002), 'Forestry cultivation subsistence of the Kemu People in Huijie Village and the tropical forest', in S. Yin Shaoting and F. Yoko (eds), *Rainforest and Rubber Forest*, Kunming: Yunnan Education Publishing House: 15–57.

Zhang, S. (2002), 'Development and changes in Menglongba of Xishuangbanna: Influence of planting rubber trees to the local economy', in S. Yin Shaoting and F. Yoko (eds), *Rainforest and Rubber Forest*, Kunming: Yunnan Education Publishing House: 244–259.

Zheng, H. (2002), 'Influence of planting rubber trees to settlements in Xishuangbanna Dai Nationality Autonomous Prefecture', in S. Yin Shaoting and F. Yoko (eds), *Rainforest and Rubber Forest*, Kunming: Yunnan Education Publishing House: 58–77.

9
Impacts of Population Migration on Land Degradation in the Tarim River Basin

Jiang Leiwen, Tong Yufen, Zhao Zhijie, Li Tianhong, Liao Jianhua and Han Wei

Land degradation is one of the most serious environmental and socioeconomic problems in the world today. More than 250 million people in over 110 countries currently are directly affected by desertification, and about one billion are at risk. Around one-sixth of the world's population is living in arid and semi-arid regions (World Bank 1999). The major arid regions are those whose population growth rates are high, while socio-development levels are low (UNDP 1999). While it has been accepted worldwide that land degradation presents the gravest challenge for those regions to develop economically and improve human wellbeing, there is no general consensus as to its underlying causes (Deng 1999). Since the 1980s, the issue of land degradation has increasingly attracted the attention of the international community, with the main focus of interest on the natural environment and ecology. Since the 1990s, the relationship between land degradation and population growth has received increasing attention, and has alarmed politicians and scientists (Knerr 2001). However, the question as to what extent population growth affects land degradation is contentious, creating confusion in policies and management programs intended to help many of the world's poorest people.

The situation of a lack of consensus on the impact of population growth on land degradation is partly due to the fact that land degradation is caused by a range of complicated factors. Demographic issues is one of the driving forces behind land degradation but is always combined with other factors in a synergetic and multi-directional manner. Another important reason behind this lack of agreement is that the majority of existing studies on the subject conventionally use population size, fertility or population density as the unit of analysis, ignoring the importance of a more significant variable—population migration. This is mainly due to the problem of data limitation, as data on migration is always less accessible and reliable than that on population size or fertility (Knerr 2001). Moreover, the majority of migration occurs within administrative boundaries and is therefore invisible in cross-national or regional statistical analysis (Geist and Lambin 2001).

In more recent years, there are increasing numbers of studies investigating the importance of population migration as both the cause and consequence of environmental degradation. On the one hand, many studies are devoted to the analysis of forced migration ('environmental refugee' or 'displaced population') due to environmental degradation (e.g. Lonergan 1998; Bates 2002; Hugo 1996; Leighton 1999); on the other hand, others discuss the impacts of migration on environment degradation. For example, a study on the driving forces of deforestation in the Brazilian Amazon indicates that population density does not have a significant effect on deforestation when other variables were taken into account in the statistical model; however, the small number of newly migrating population to the Amazon have significantly more impact than those already living in the area (Pfaff 1997). A systematic analysis of 152 case studies of tropical deforestation by Geist and Lambin (2001) also concludes that population does influence deforestation rates, but not primarily because of its sheer increase due to high fertility rates; the influx of migrants into forested areas with low population density is the more important demographic variable in explaining deforestation. A synthetic report on this issue has recently been provided by Richard Bilsborrow (2002). The major areas of study focus on tropical forests in Latin America and Asia, for example studies on the Brazilian Amazon (Mahar 1979; Moran 1981; Hecht 1985; Hecht and Cockburn 1990; Schmink and Wood 1993), Ecuador (Pichon 1997; Pichon and Bilsborrow 1999; FAO, 1997 and 2001), Panama (Heckandon and Mckay 1984; Joly 1989; Schelhas 1996) and on Indonesia and Malaysia (Oberai 1998; Heyzer 1995). Others study the impact of migration on desiccation in dryland areas of sub-Saharan Africa and Latin America, focusing, for example, on Tanzania (Darkoh 1982), Sudan (Ibrahim 1987; Little, 1987; Bilsborrow and Delargy 1991), Egypt (Findlay 1996), and the Yucatan Peninsula, Mexico (Ericson et al. 1999). Although those case studies provide a good body of knowledge for generalization, theory and methodology on the environmental consequences of migration are still very limited (Bilsborrow 2002).

Our study is designed to explore the interactions and their operational mechanisms between migration and land degradation in Tarim River, Xinjiang of China. In this paper, exploiting a multi-disciplinary approach, the historical process of demographic dynamics, land-cover and land-use changes across the period of 1950–2000 are firstly inspected. Secondly, the extent and proximate causes of land degradation are analyzed. The impact of population growth, migration in particular, on land degradation is then carefully studied and simulated. Finally, conclusions and policy recommendations for the local population and economically and environmentally sustainable development are provided.

Background

Xinjiang Uygur Autonomous Region in northwestern China has an area of 1.66 million km^2 and accounts for about one-sixth of the total land area of China (Figure 9.1). It is an arid area with a fragile and vulnerable ecosystem, subject to irreversible desertification. Due to the disparities in ecological conditions and socioeconomic development levels, population density differs markedly between the northwest and southeast of China (Hu 1936). Since the foundation of the People's Republic in 1949, the Chinese government has directed large waves of migration from the east to the west in order to enhance socioeconomic development levels and to explore land and mineral resources in the west. However, the unbalanced population distribution has not changed significantly in the past decades, while a large volume of migrants together with reckless land reclamation, resource exploration, and improper water use have induced serious land degradation problems in the west.

In the government-directed population movements of past decades, Xinjiang was the major migrant-receiver. Migration in Xinjiang can be divided into five stages (Figure 9.2):

1. The first tide of in-migration (1949–1961): After the foundation of the People's Republic of China, large numbers of military troops from the east settled and established 'Production and Construction Troops' in Xinjiang. Thereafter, many demobilized soldiers, intellectuals and farmers were either organized by the government or attracted by the relatively high salary and better employment opportunities, and moved to Xinjiang. Particularly in the years 1959 and 1961 when China experienced three years of severe economic difficulties, large volumes of in-migrants from the east sought food and jobs in Xinjiang, which resulted in an historical record of high in-migration rates. During the first period of twelve years, the population of Xinjiang increased from 4.33 million in 1949 to 7.1 million in 1961, among which net migrants accounted for 67.8%.
2. The dramatic out-migration wave (1962–1963): After the Three-Difficult-Years, the economic situation in the east improved. More importantly, the rapid population growth generated social and economic problems for Xinjiang and led to a 'policy of simplification'—many newcomers to Xinjiang were sent back to their origins. In 1962–1963, Xinjiang experienced a significant negative net migration.
3. The continuous net in-migration period (1964–1980): The simplification policy did not last long. Xinjiang continuously received net in-migrants afterwards, although the volume of migrants decreased gradually. In the

Figure 9.1 Map of China

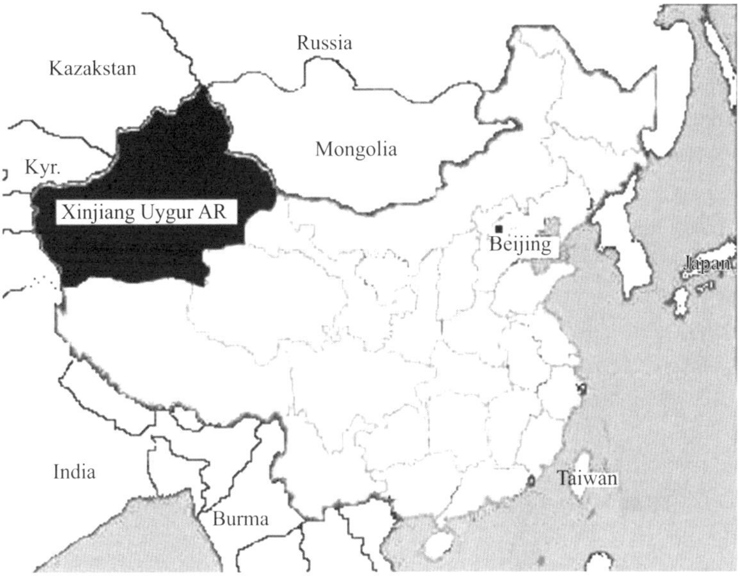

period of 1964–1980, the population of Xinjiang increased from 7.44 million to 12.83 million, of which migration contributed 27.8%.

4. The net out-migration period (1980–1989): Since the introduction of economic reform and the open policy in 1979, economic development policy in China shifted from an equilibrium development strategy to the 'east coastal area development strategy'. Many previous in-migrants chose to move back to their places of origin in the east. In the 1980s, Xinjiang lost 245,700 people in net out-migration.
5. New in-migrating tide (1990–2000): At the end of 1980s, China experienced economic recession. A significant floating population who moved for employment opportunities without change of their household registration were driven out from the initial main migrant receiving areas in the east and turned to seek chances in the western regions. More importantly, the central government, facing enlarged socioeconomic development gaps between the east and west, started the new west development strategy in the late 1990s. As one of the main regions of new investment, Xinjiang established a number of projects of energy exploration, infrastructure construction, and environmental rehabilitation, and received an increasing number of in-migrants. The annual net in-migrants to Xinjiang increased from 44,800 in 1990–1995 to 75,700 in 1996–2000. In the 1990s,

Figure 9.2 Population changes in Xinjiang, 1950–2000

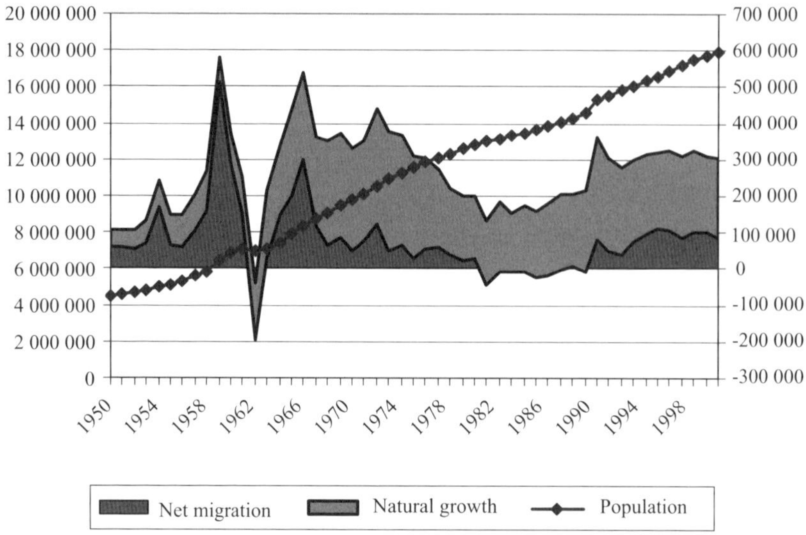

the population of Xinjiang increased by 2.64 million, among which migration accounted for 18.3%.

Therefore, before the 1970s, migration was the main force of population growth in Xinjiang, while natural growth became more important at a later stage as migration rates decreased and stabilized (Figure 9.3).

It should be noted that the above-mentioned population growth only refers to permanent residents, while the unregistered floating population was not taken into account. Actually, there is an increasing number of migrants to Xinjiang moving without official changes of household registration and being regarded as a floating population since the economic reforms of the early 1980s. The official statistics indicate that the number of the floating population has increased rapidly since 1983 (Figure 9.4). Moreover, it is believed that the official statistics are seriously underreported regarding this population. For example, the official registered floating population as of 1990 was only 330,200, compared with 710,600 reported by the 1990 Census (Yang et al. 1996). Figure 9.4 also shows that the distribution of the floating population significantly changed between sub-regions in Xinjiang: the proportion in southern Xinjiang increased from about thirty percent in 1983 to approximately fifty percent in 1993. This increase is mainly due to the large land reclamation in the Tarim River Basin.

Figure 9.3 Population dynamics of Xinjiang, 1950–2000

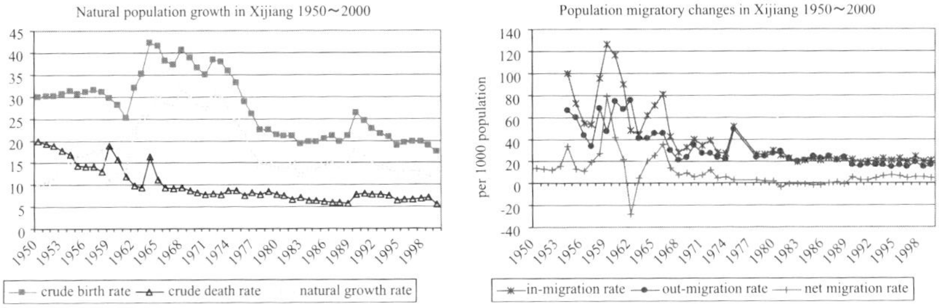

Figure 9.4 Floating population of Xinjiang, 1983–2000

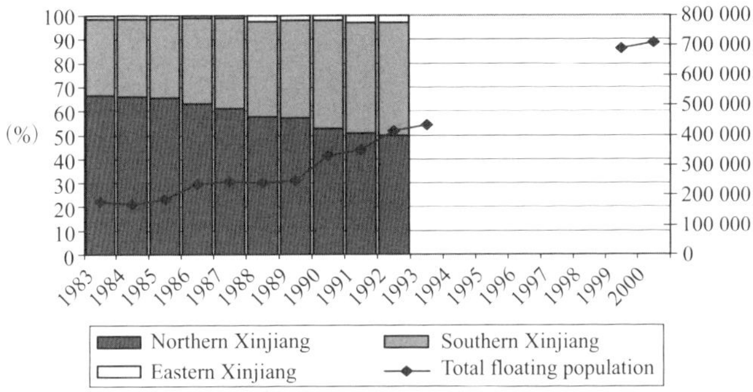

In the past five decades, the population of Xinjiang increased more than fourfold, with about thirty percent of this increase due to migration. Accordingly, the population density of Xinjiang increased from 2.7 persons/km^2 in 1949 to 11.2 persons/km^2 in 1999. Although it is at a lower rate than the national average, ninety percent of Xinjiang's population is concentrated on a ridge of mountains between two areas of desert, which only accounts for 4.5% of its total land area. Population density in those oases in 1999 was as high as 260 persons/km^2, which is almost two times that of the average population density of China. Rapid population growth generated great pressure on the fragile ecosystem, and induced enormous land degradation. The desert area increased by 3784 km^2 in the period from 1949 to 1994—a rate of 84.2 km^2 annually—and currently amounts to over 430,400 km^2. Forest coverage has dropped to as low as 1.68%, which is only one-eighth of the national average. Over-grazing significantly decreased the productivity of grassland, and worsened its savanna ecosystem. Overuse of fertilizers and pesticides induced land pollution, and caused a decrease in the average content of organic matter in the soil to less than two percent in general.

Land degradation has become the most serious environmental problem in this region, causing stress, morbidity, and decreased wellbeing amongst the affected population. The Tarim River Basin is one of the areas in Xinjiang which has experienced the most rapid population growth, the biggest changes in land cover and land use, and is encountering the most severe problems of land degradation. Therefore, our study in this chapter will focus on the Tarim River Basin.

Migration and land degradation in the Tarim River Basin

Changes of population migration and cultivated land

Xinjiang is divided into two major sections by the Tianshan Range (Figure 9.5). The northern part is the Jungarian Basin, while the southern part is the Taklimakan Desert and the Tarim Basin. Located at the heart of the Eurasian Continent and being surrounded by mountains, the Tarim Basin is one of the driest areas in the world. Its annual precipitation is only fifteen to sixty-five mm, while the annual evaporation is around 2000–3000 mm. The ratio of annual precipitation to the potential evaporation is only 0.014 to 0.02 (Ji 2001). Conversely, the Basin is rich in heat and light, with annual solar radiation of 5.8 billion J/m^2. This would be ideal for irrigation agriculture, if water resources were available (Zhou et al. 2000).

The limited amount of precipitation in the area is mostly in the form of snow and concentrates on mountain slopes. Small water streams generating from the snowmelt on the south slope of Tianshan Mountains converge and eventually form China's longest in-land River—Tarim River. In the local Uygur language, *tarim* means 'converging of waters' (World Wildlife Fund, 2001). Running through China's largest desert, the Taklimakan, the Tarim River waters a strip of Hutong poplar forest which stabilizes the sand and moderates the climate.

Before 600 AD, the Tarim River flowed naturally and swung cross the desert from year to year watering a forty to 100 km wide belt of forest. Except the loss due to evaporation and seepage in the middle reaches, all water flowed along the river into Luobupuo Lake in the lower reaches, where the civilization of Luolan emerged. Luolan was ever one of the biggest towns along the ancient silkroad. Due to natural growth and in-migration, population pressure on cultivated land drove people out to explore new territory. Meanwhile the advancement of productive tools and hydrological skills enabled people to draw water from the river and irrigate the newly

Figure 9.5 Geographic conditions of Xinjiang and Tarim Basin

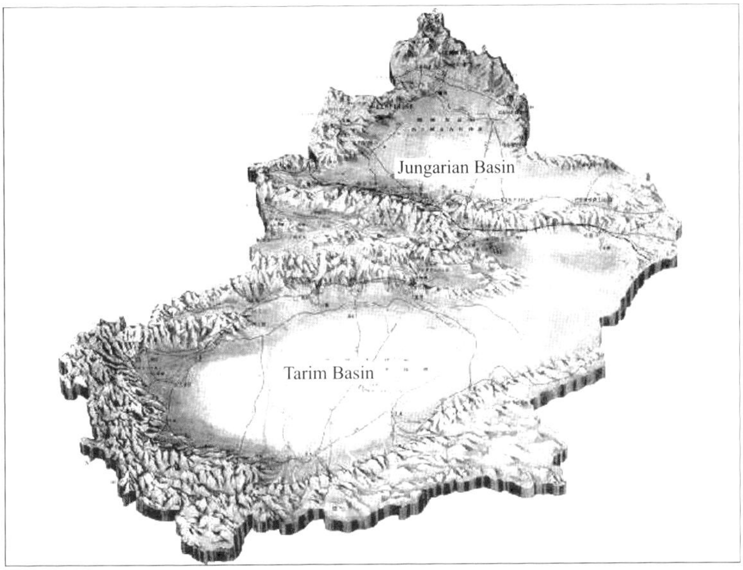

reclaimed land in the upper reaches. As increasing numbers of oases were constructed in the upper reaches, water reaching the downstream declined, which eventually led to the extinction of Luolan City.

Although the ecosystem in the lower reaches of Tarim River had deteriorated since 600 AD, the most significant change only happened after the early 1950s. Since the foundation of the People's Republic, the Tarim River Basin has become one of the most important cultivation bases. Besides land reclamation by the local community, since 1950 Xinjiang Productive and Construction Troops have established Division 1 in the upper reaches and Division 2 in the lower and middle reaches of the Basin. From then, large numbers of demobilized soldiers, intellectual youth and peasants from the east have moved into the area and constructed more than seventy reservoirs to irrigate the reclaimed cultivated land. The total population of these two divisions increased from less than 20,000 in 1950 to 440,000 at the end of the 1970s, while the total cultivated land jumped from around 20,000 ha in the early 1950s to 142,000 ha at the end of that decade (Figure 9.6). Although the amount of cultivated land in Division 2 stabilized since the 1960s before slightly increasing again in the 1990s, it continued increasing in Division 1 in the upper reaches. The stagnation of land reclamation in Division 2 after the 1960s was mainly due to the limitation of water resources, given that no land could be cultivated without irrigation. The total population decreased in the 1980s due to the previous in-migrants returning to the east after the introduction of China's economic reforms and the Coastal Development Strategy. However, the population decreased more significantly in Division 2 than in Division 1, since the latter is privileged in terms of water accessibility and consequently economic benefits.

With the growth of population and cultivated land, the total sown areas in the two divisions increased continuously (Figure 9.7). The increment of sown area was dramatic in the 1950s and 1960s, stagnated in the 1970s and 1980s but resumed since the 1990s. Composition of the sown areas also changed over time. Before the 1980s, most of the agricultural land was under grain crops before the 1980s, given that China implemented an agricultural policy of 'Grain as the key' in order to meet the substantive food requirements of the huge population. As economic interests became the main consideration after the economic reforms of the 1980s, cash crops with higher economic reward were more commonly planted, and gradually became the major agricultural products in the Tarim River Basin. In the period from 1966 to 1985, rice accounted for more than twenty percent of the total sown area. However, when cultivated land was increasingly engaged in cash crop plantation, the proportion of area under rice was reduced. More importantly, the decline in rice plantation was mainly due to water constraints, since rice is one of the most water-intensive crops. Facing water scarcity, many cultivated lands had to be switched to plantations of less water-consuming species. Further, the decrease in rice plantations was much more significant in Division 2 than in Division 1, given the serious downstream water shortage.

Most of the reduced sown area under grain crops was used for cotton plantations. Xinjiang has a global reputation for producing high quality cotton—long fiber cotton. Moreover, cotton plantation is relatively less water intensive. Since the 1980s, China has become one of the biggest textile and clothing exporters in the world, causing a rapid increase in the price of cotton. Driven by market demand, large amounts of cultivated land in the Tarim River Basin were transferred into cotton production, and a large area of unused land was also reclaimed. The proportion of cotton fields to the total sown area dramatically increased from about ten percent in both divisions in 1980 to about forty-five percent in Division 2 and to above sixty-five percent in Division 1 in 2000.

Meanwhile, Xinjiang started to implement a decentralization policy, allowing local government to assign unused land for reclamation in order to increase income and generate revenue for said government. Combining with the household production responsibility system, this policy provided economic incentives for the local government and people to reclaim more

Figure 9.6 Population and cultivated land changes in Divisions 1 and 2

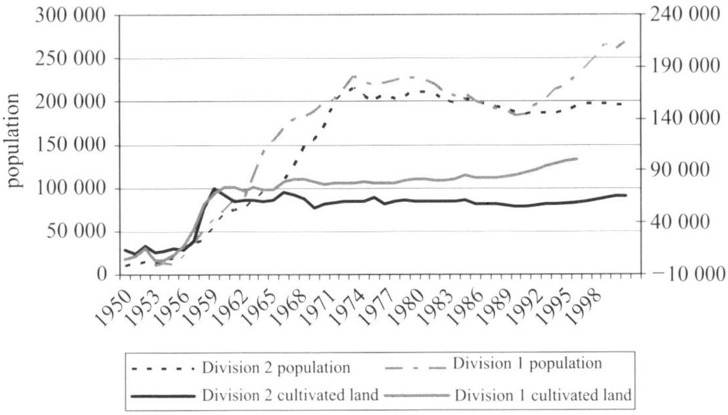

Figure 9.7a Sown area and rice and cotton plantations in Division 1

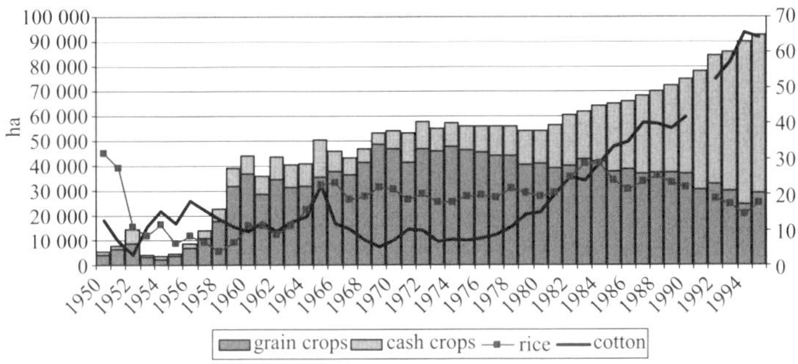

Figure 9.7b Sown area and rice and cotton plantations in Division 2

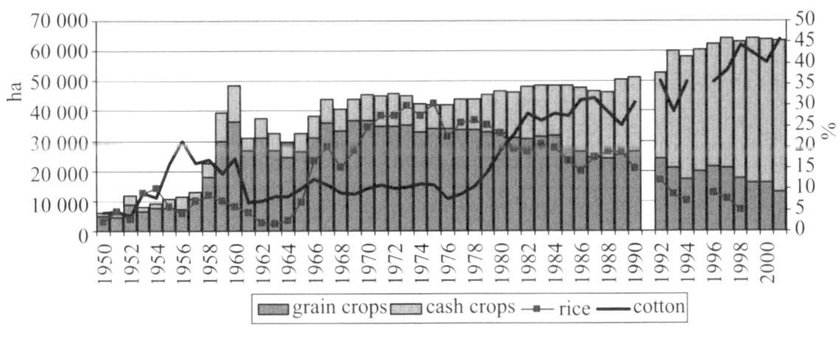

land. When forest and pastures were used up, the unfavorable land with higher salt content was also reclaimed. A large volume of water from the river was drawn to flush the salt content of the newly reclaimed land, and was then channeled back to the river. The more land reclaimed, the less volume and poorer quality of river water flowing downstream. As a result, cultivation in Division 2 was seriously constrained by the water quantity and quality, although they had the same anxiety of land reclamation as Division 1. The total area of cultivated land significantly increased in Division 1, but only slightly increased in Division 2. Therefore, the total cultivated land of Division 1 exceeded that of Division 2 by more than forty percent in 1995, although there was more cultivated land in Division 2 than in Division 1 before the early 1960s. The change of population size between the two divisions followed a similar pattern. The officially registered population size of Division 2 was larger than Division 1 before the early 1960s. However, in the year 2000, the officially registered population of Division 1 outnumbered that of Division 2 by thirty percent. Moreover, when the floating population is included, the differences are even larger.

Due to the scale of land reclamation and the expansion of cotton plantations a large inflow of itinerant migrants moved into the Tarim River Basin. The statistics of the floating population in Division 2 was about five to eight percent of its total permanent population in the period from 1990 to 2000. The statistics for Division 1 are not available. However, an estimate can be made based on the labor demands of cotton production. Cotton plantations are labor intensive. Most of the cotton fields in the Tarim River Basin are contracted to migrants moving from the provinces of Sichuan, Henan and Gansu. Provided that one laborer manages one to three ha of cotton before it reaches the harvest stage, the reported cotton field of 63,600 ha in Division 1 in 1995 would need 20,000 to 64,000 laborers for prior-harvest management. Considering their accompaniers, the floating population before harvest is estimated at 30,000 to more than 100,000 in Division 1, equal to ten to thirty-five percent of its permanent population. Moreover, cotton harvest is the most labor-intensive period. In general, one laborer could harvest ten to fifteen kg of cotton a day on average, and one hectare of cotton which produces 3000 kilogram on average requires 200–300 labor days. Therefore, to harvest the total cotton fields of Division 1 and Division 2 in 1995 requires seventeen to twenty-six million labor days. If we assume each laborer spends sixty days working in the cotton field, the cotton harvest would require about 280,000 to 440,000 laborers, of which in-migrants are the majority. It is estimated that seventy-five percent of the migrants moved to Division 1. Therefore, the floating population size in Division 1 is much bigger than that in Division 2.

Among the floating population, those who come for cotton harvest often stay for one to three months only, while those who engage in cotton field management also face uncertainty given their household registration status. Therefore, the Tarim River Basin is only a temporary abode for them. It is observed that the floating population in the Tarim River Basin extracted water, land and wood resources much more recklessly than the local people, generating destructive effects and causing serious land degradation. This phenomenon is also reported by studies of new or seasonal migrants in tropical forest frontiers, indicating that the environmental impact of a small number of migrants would significantly exceed that of the same number of already settled people (Geist and Lambin 2001).

Worsening water resources

Large scale in-migrants and land reclamation in the Tarim River Basin induced heavy pressures on water resources.

The Tarim River system consists of three tributaries—the Aksu River, Yarkant River and Hotan River. The whole basin can be divided into four parts: 1) source region—the catchment areas of the three tributaries; 2) the upper reaches—from Aral to Yingbazha; 3) the middle reaches—from Yingbazha to Qara; and 4) the lower reaches—down from Qara. The total volume of water resources from the tributes is relatively fixed: 7.5 billion m^3 from Aksu River, 7.4 billion m^3 from Yarkant River and 4.5 billion m^3 from Hotan River, which amounts to a total water source of 19.5 billion m^3. However, in the period of 1957 to 1995, water resources reaching the river trunk decreased from 5.62 billion m^3 down to 3.94 billion m^3 (Table 9.1). An increasing amount of water was consumed in the source regions. For example, statistics in Aksu River, which provides more than seventy-five percent of the water source of the Tarim River trunk, shows that the volume of water flowing out of the Aksu region decreased by thirty percent in the 1980s to 1990s. From the early 1950s, Division 1 established the Aral and Jingsezi Cultivation Bases in the Yarkant and Aksu River areas. More than forty reservoirs of various sizes with a capacity of 1.6 billion m^3 were built in Yarkant River. Due to the very high evaporation rate, more than fifty percent of the water of the reservoir was evaporated, while the poor irrigation facilities only produced a 0.35 to 0.4 canal effective utilization coefficient. Moreover, the groundwater which is usually one of the most important irrigation sources in the arid areas has only been exploited by three percent. Furthermore, the low water price (0.006 RMB *yuan*, or 0.07 US cent per m^3) gives no economic incentives for people in the source regions to regulate their water use. Water consumption of each hectare of land in the source region was as high as 15,000 to 22,500 m^3. As a result, almost all the run-off

Table 9.1 Changes in water consumption in the Tarim River Basin

	Run-off volume at Aral	Water consumed (100 million m^3)			Water consumed (%)		
Time period	100 million m^3	Upper	Middle	Lower	Upper	Middle	Lower
1957–1959	56.2	14.8	25.8	15.6	26.4	45.9	27.7
1960–1969	53.2	16.4	24.4	12.4	30.8	45.9	23.2
1970–1979	43.2	15.0	22.1	6.7	34.2	50.6	15.2
1980–1989	44.8	18.2	22.7	3.9	40.6	50.7	8.7
1990–1995	39.4	17.1	19.2	3.1	43.5	48.7	7.8
Average	46.5	16.3	22.5	7.7	35.1	48.4	16.5

Source: Ji (2001: 144)

Table 9.2 Changes in water volume of the Aksu River

Year	Water flow-in	Water drained	Water to Tarim River
1980	73.5	45.6	33.7
1981	81.5	41.7	46.5
1982	69.1	43.0	31.3
1983	72.7	44.2	34.8
1984	75.7	41.9	33.0
1985	62.4	41.8	17.3
1986	71.2	47.0	35.4
1987	75.6	42.4	40.9
1988	75.6	47.3	38.4
1989	68.5	56.1	29.2
1990	78.1	57.5	29.2
1991	75.7	58.0	20.1
1992	69.9	57.4	26.9
1993	69.2	59.7	25.2
1994	96.3	61.0	37.1
1995	93.7	64.4	37.8
average	75.5	50.6	32.7

Source: Song et al. (2000: 53)

of Yarkant River was used up in situ. Since the middle 1970s, the only water flowing from the Yarkant River into the Tarim is used water after irrigation or after flushing the salt content from reclaimed land.

In the period from 1950 to 2000, the cultivated land in the source region almost doubled. The large scale of previous reclamation had already exhausted land resources suitable to cultivation. Remaining unused land are second or third degree lands with high salt content. Division 1 in the Aksu River pledged to 'reconstruct a new Division 1', and engaged in the reclamation of

Figure 9.8 Salt content changes at the Aral Hydrological Station

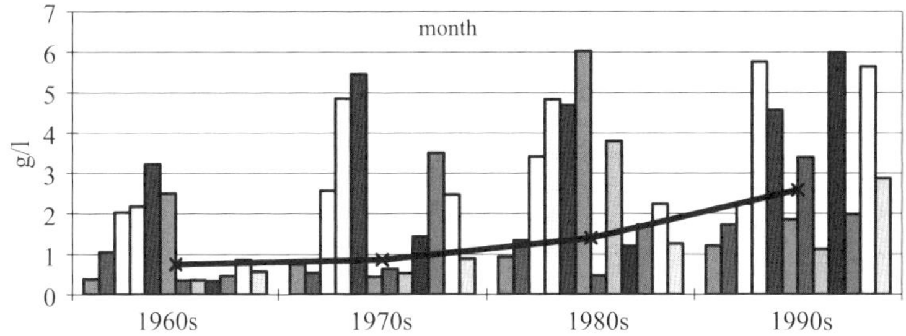

50,000 ha in the period 1995–2000, while Aksu local government was trying to reclaim another 67,000 ha at the same time. In order to enable the total reclaimed land to be cultivable, sixty-seven million tons of salt would need to be flushed out. If only half of the salt is discharged into the Tarim River, the annual salt content of the river will increase by 6.7 million tons, which will seriously deteriorated its water quality. According to the monitoring in the Aral Hydrological Station, the monthly salt content of the Tarim River significantly increased in the period from the 1960s to the 1990s (Figure 9.8). The annual average salt content of the river increased from 0.75 g/L in the 1960s, to 0.86 g/L in the 1970s, 1.4 g/L in the 1980s and 2.6 g/L in the 1990s. In the past ten years, the annual increase of water salt content in the main stream of the Tarim River amounted to five percent (Zhou et al. 2001).

In recent years, people in the middle reaches of the Tarim River have also engaged in land reclamation and ecological construction projects which have increased water consumption. As a result, when reaching the downstream areas, the water is small in volume and of poor quality. Statistics shows that the water volume flowing through Qara Hydrological Station decreased from 1.56 billion m^3 at the end of the 1950s to only 0.28 billion m^3 in 1995, a reduction of 560%. China's ever largest lake Luobupo in the lower reaches of the Tarim River was eventually dried up and buried under the desert in the late 1960s. The other once large lake Taitema lost its water supply leaving only a dry lake basin by the early 1980s. The Tarim River now ends at Daxihezi Reservoir and has been shortened by 320 kilometers compared to its length in the early 1960s.

Land degradation

Historically, the Tarim River Basin is rich in terms of biodiversity. The swinging streams of the Tarim River watered a large range of plants and were home to various flora and fauna, including *Populus euphtatica*,

E. angustifolia, *Pantera Tigris*, *Cervus elaphus*, *Gazella subguturosa*, *Aspiorhynchus laliceps* and *Ciconianigra*. Flood seepage provides seventy-five percent of the source of groundwater which feeds the forest, bush and meadows. However, while seventy to eighty percent of the water source in the Tarim River Basin was controlled by the irrigation system and human settlement, ecological water was marginalized. As a result, the landscape of the Tarim Basin was greatly altered.

Exploiting the aero-photographic imageries, the evolution of land-cover and land-use changes in different areas of the Tarim River Basin can be analyzed over time. Scientists from Xinjiang Institute of Ecology and Geography invested great effort on this study. Choosing the areas surrounding the Aral in the upper reaches, Yingbazha in the middle reaches, and Alagan in the downstream area as the study sites, the aero-photographic imageries of these three locations in 1959, 1983, 1992 and 1996 were obtained and digitized. After a series of works of categorization, map-editing and ground-truthing, the natures of land cover and land change in the three areas at the four time points were characterized and calibrated.

Analysis on the Aral area in the upper reaches indicates that cultivated land expanded the most in the period of 1959–1996, at the expense of the Hutong Poplar forest, pastures, and bush (Table 9.3). With the construction of the large scale Shengli Reservoir, the natural water area was substituted by and transformed into a human-made water zone. Although built-up areas were still relatively insignificant, the land used for residential purpose increased about threefold over less than four decades. Desertification is not a problem in this area. In fact, the desert area decreased slightly in all periods of the forty years. However, the Aral suffered serious problems of secondary salinization, due to the ineffective drainage system and excessive water use. It is estimated that about thirty-five to forty percent of the cultivated land in the Aral area are suffering salinization problems. By 1983, the abandoned cultivated land due to salinization accounted for about five percent of the total cultivated land.

In the middle reaches of Yingbazha area, the landscape was also greatly changed by in-migrants. In particular, Santamu Forestry Station was established for exploiting forestry resources in 1972. In the eight years from 1972–1980, 26,403 m^3 of Hutong poplars was cleared. Almost at the same time, the establishment of the Liberation Canal Farm caused a rapid increase of land reclamation and degradation. From 1960 to 1983, the total area of desert increased from 917.36 km^2 to 1065.27 km^2; the proportion of desert to the total area increase by about ten percentage points (Table 9.2). This rapid process of desertification was reversed in the 1980s when

Table 9.3 Land cover and land use change in the Aral area (hm^2)

Land category	1959	1983	1992	1996
Total area of the imagery	156 888.0	156 888.0	156 888.0	156 888.0
Cultivated land	21 128.4	29 808.3	34 025.5	37 924.4
Hutong poplar forest	14 054.0	12 957.4	11 183.3	10 388.6
Bush	24 378.5	18 571.9	17 073.9	16 146.3
Desert pasture	19 836.5	20,052.6	18 782.0	18 635.4
Pasture	29 401.6	26 277.4	26 806.9	23 443.2
Residential area	615.3	942.5	1 698.6	1 829.6
River or riparian area	8 087.1	6 427.2	6 665.3	7 514.9
Lake	1 143.7	150.3	146.0	111.9
Reservoir	0.0	4 707.0	4 707.0	4 707.0
Desert	36 242.9	35 654.3	35 338.2	35 235.1
abandoned cultivated land	0.0	1 339.1	474.3	951.6

Source: Ji (2001: 159)

Table 9.4 Annual desertification rate of Yingbazha area (%)

Year	Extremely strong desertification	Strong desertification	Medium desertification	Slightly desertification	Sum	% desert to total land area
1960–1978	8.1	1.72	0.44	0.39	0.76	59.07
1978–1983	7.17	10.57	3.16	-2.56	0.27	67.67
1983–1990	1.31	1.92	0.37	-2.72	-1.12	63.36
1990–1992	1.84	-0.32	-1.49	0.07	-0.14	63.18

Source: Ji (2001: 160, 381)

several natural protection projects were implemented. The natural vegetation was slightly recovered and improved as a result. Unfortunately, the natural vegetation recovering effort ended in the early 1990s, when Yingbazha was designated as one of the important bases of the Tarim Basin oil exploration program. After the opening of the Luntai-Minfeng Desert Highway in 1995, Yingbazha has attracted a new wave of in-migrants who began a new round of land reclamation and degradation.

Land degradation was most serious in the Alagan area in the downstream reaches of the Basin. Because river water was drained upstream, the desert to the total area in Alagan was already 86.98% in 1959 (Table 9.5). This situation worsened when more migrants came, more land was cultivated, and more water was used for irrigation in the upper and middle reaches of the river basin. The proportion of desert in Alagan reached 94.78% in 1996. Because there was no river water reaching this area by the early 1980s, the

Table 9.5 Desertification of the Alagan area

Year	Total		Extremely strong desertification		Strong desertification		Medium desertification		Slightly desertification	
	km^2	%	km^2	%	km^2	%	km^2	%	km^2	%
1959	1371.22	86.98	476.15	30.20	95.78	6.08	393.42	24.96	405.87	25.74
1983	1460.29	92.63	497.63	31.56	96.58	6.13	454.95	28.86	411.13	26.08
1992	1487.26	94.34	532.24	33.76	145.01	9.20	438.01	27.78	372.00	23.60
1996	1494.29	94.78	555.41	35.23	184.00	11.67	437.91	27.78	316.97	20.10

Source: Song et al. (2000: 268)

table of groundwater decreased from seven m in 1973 to 10.40 m in 1989 and 12.65 m in 1997 (Ji 2001). Consequently, plants, animals as well as humans were almost completely driven from this area in the late 1990s.

Quantifying the impact of population changes on land degradation

Concluding from the analysis above, although the upper and middle reaches of the river basin had different experiences in land-use and land cover change, the problems of land degradation caused by rapid in-migration, large scale land reclamation, excessive water consumption and pollution in the whole river basin cumulated in the lower reaches. Therefore, we pay special attention in this section to the situation of land degradation in Yuli County which covers most of the lower reaches of the Tarim River Basin.

Being surrounded by the Takelamagan Desert from the southwest and Kuluke Desert from the northeast, Yuli is one of the biggest counties in China. In the early 1950s, the area was only inhabited by around 10,000 Uygur and Mongolian nomads. Division 2 of Xinjiang Production and Construction Troops established nine corps (later reformed into five corps) from Qara to Tieganlike along the Tarim River in 1957–1958. Large numbers of in-migrants quickly doubled the Yuli population (Figure 9.10), and transformed forest, bush, pasture and water areas into cultivated land and irrigation systems. By the end of the 1960s, the construction troops had a total cultivated land of 267,000 km^2, and were named among China's 100 best agricultural enterprises. However, facing water scarcity from the middle 1970s, the expansion of population and cultivated land stopped. To solve the problem of water shortage, Yuli constructed Kuta Canal, carrying water from Kongque River into the areas of the corps. When the river water from the upper reaches of Tarim decreased further, Kuta Canal could hardly help to compensate the water loss. The corps was forced to abandon about forty-five percent of its total cultivated land in the 1970s–1980s. Moreover,

Figure 9.9 Population change in Yuli County

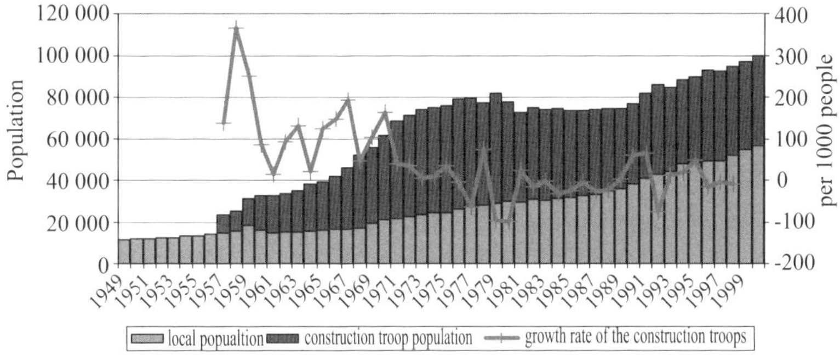

among the remaining 151,000 km^2 cultivated land, the actual sown area was only 133,000 km^2 according to the statistics of 1995. Given the serious land degradation problems, population growth in the construction troops was reversed since the 1980s due to negative net migration (Figure 9.10).

In order to quantitatively study the problem of land degradation and its interaction with population changes, we obtained for Yuli the aero-photographic imageries of 1958 and remote sensing data of 1999. Adopting the 3-S methods and technologies, we analyzed and compared the conditions of land use and land cover at 1958 and 1999. The results of our analysis are given in Table 9.6, and figures 9.11a and 9.11b. It is evident that land use and land cover has significantly changed over the past forty years. For the whole county, the wasteland increased from 90.8% to 95.1%. Although other types of land use only account for a small proportion, their changes are very impressive: the water areas, woodlands and pasture shrunk by half, while the built areas increased by three times and cultivated land grew fivefold.

Moreover, the comparison of Yuli County as whole does not capture the major landscape changes induced by human intervention. Since population increase only happened in the areas able to provide basic living conditions, changes in the vast desert are limited. Therefore, changes in land use and land cover in the area under the jurisdiction of construction corps along the Tarim River would more accurately reflect the interactions between land degradation and population activities. Analysis of changes of land characteristics in the areas of construction troops gives a much stronger impression of land degradation over time: wasteland increased by ten percent, cultivated land and built-up areas increased by seven percent and other types of land altogether decreased by seventeen percent. The statistics clearly conveys that the downstream of Tarim River Basin experienced critical land degradation in the period of 1958–1999.

Table 9.6 Land cover and land use changes in Yuli County, 1958 and 1999

	Total				Construction corps			
	1958		1999		1958		1999	
	km^2	%	km^2	%	km^2	%	km^2	%
Water area	539.45	0.92	229.49	0.39	151.56	6.30	54.81	2.15
Wasteland	53414.31	90.85	55918.84	95.11	1565.68	65.12	1921.12	75.48
desert	50622.23	86.10	51601.72	87.76	1552.06	64.55	1827.44	71.80
saline area	2100.21	3.57	3575.63	6.08	13.58	0.56	77.62	3.05
marsh	691.88	1.18	741.49	1.26	0.00	0.00	16.06	0.63
Cultivated land	59.82	0.10	305.72	0.52	38.07	1.58	206.01	8.09
Woodland	2613.02	4.44	1328.39	2.26	465.28	19.35	228.90	8.99
forest	1709.61	2.91	1051.52	1.79	170.06	7.07	152.87	6.01
bush	903.41	1.54	276.87	0.47	295.22	12.28	76.03	2.99
Pasture	2161.93	3.68	988.31	1.68	180.40	7.50	118.73	4.66
Built-up area	8.36	0.01	26.15	0.04	3.46	0.14	15.72	0.62
Total	58796.89	100.00	58796.89	100.00	2404.45	100.00	2545.30	100.00

In our analysis, we define cultivated land and built-area as 'human settlement'; desert, Gobi, and salina as 'desert areas'; and other types of land as 'natural arable areas'. We argue that:

1. Expansion of human settlement occurs at the expense of natural arable areas since intrusion of migrants almost exclusively occurs in natural arable areas.
2. Growth of human settlement is accompanied by desertification, because natural arable areas are often converted into desert areas by destructive human activities; and devastation of human settlements due to negative environmental responses also leads to expansion of desert areas.
3. Extension of human settlement and desert areas on the one hand, and contraction of natural arable areas on the other shorten the distance between human settlement and desert.

Therefore, in an arid area like the Tarim River Basin, migrants' intervention generally results in expansion of either human settlement or desert area. The natural arable land areas are in a transitional status and subject to substitution by either human settlement or desert areas. Figures 9.11a and 9.11b clearly demonstrate that in the period of 1958 to 1999, the corps converted woodland and pasture into cultivated land or built-up areas, and natural water bodies into man-made reservoirs and canals. While the water previously used for forestry and pasture was drained into irrigation systems, a large range of natural arable areas were degraded into desert. Losing the buffer zone of the natural arable areas, the human settlement could hardly resist the intrusion

Figure9.10a Land cover and land use of Yuli County in 1958

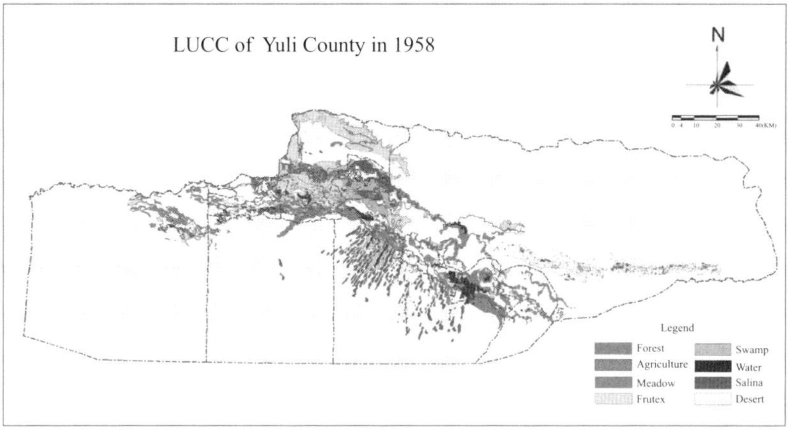

Figure 9.10b Land cover and land use of Yuli County in 1999

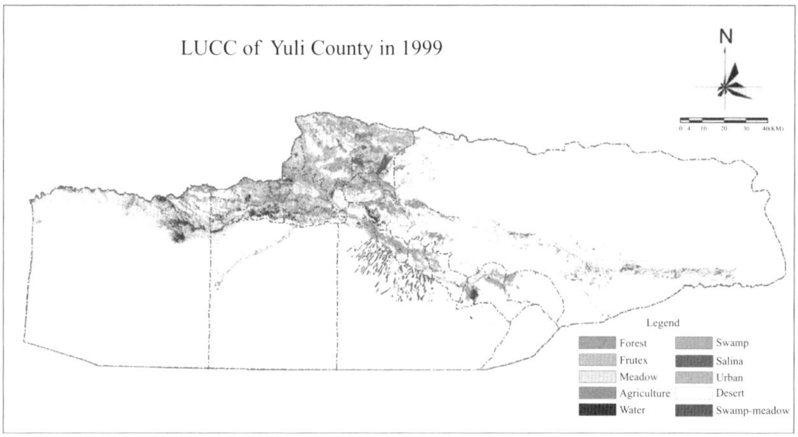

of desert. Consequently, cultivated land was abandoned and villages were buried under sand. For example, Corp 35 was forced to give up two-thirds of its cultivated land and two-fifths of its companies, facing problems of water scarcity and sand dune intrusion in the 1980s and 1990s.

To quantify the impacts of population changes on land degradation in the downstream of the Tarim River Basin, we adopt the system dynamics approach, considering the complicated interactions between interlinked factors. Although our analysis focuses on the downstream, land degradation in the downstream area is affected by population growth, human activities and water consumptions not only in situ, but also in the upper reaches.

Figure 9.11 Inter-linkage between the three main subsystems

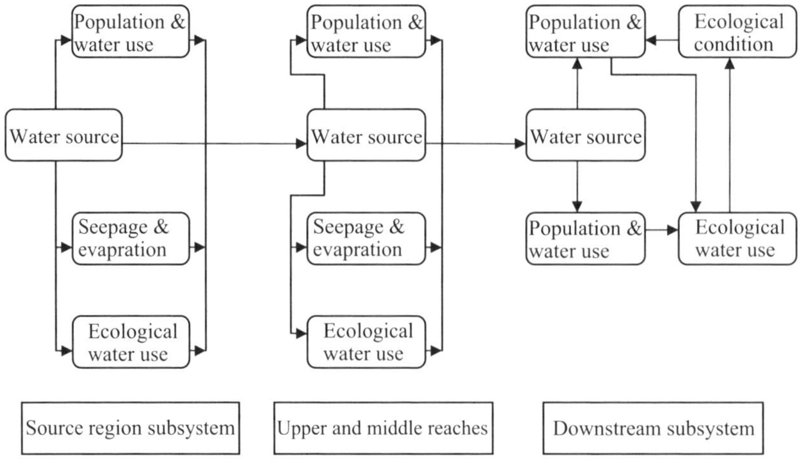

Figure 9.12 Correlation between subsystems in the downstream main subsystem

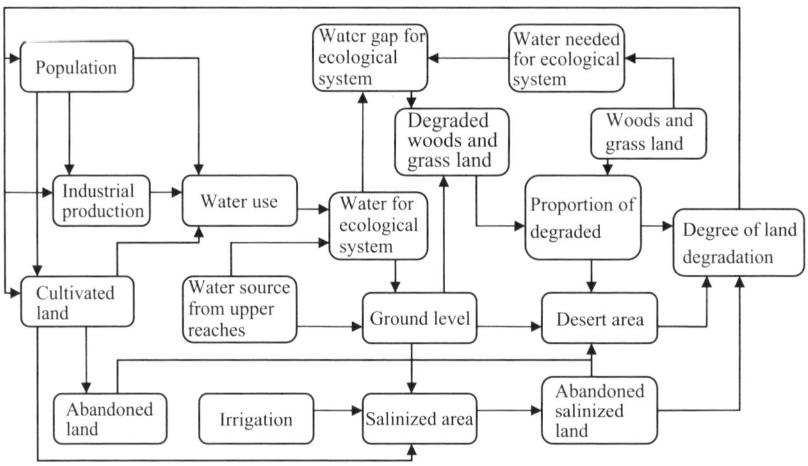

Therefore, our model is constructed taking into account the whole river basin system which comprises three main subsystems: 1) the source region subsystem, 2) the upper and middle reaches subsystem, and 3) the downstream subsystem. Each main subsystem consists of a number of subsystems, for example the population subsystem, land exploration subsystem and ecological subsystem. From source region to upper and middle reaches and then to downstream, the number of subsystems and factors as well as the complexity of the interactions increases. In our model, water flow and water consumption are the key factors connecting the main subsystems (Figure 9.12). Figure 9.13 presents the brief inter-correlations between the subsystems and factors within the downstream main subsystem. The frameworks for the other two main subsystems in the upper reaches are basically the same but simplified versions.

In the downstream main subsystem, population growth and industrial and agricultural production boosts water consumption. Given the fixed volume of water sources, increments in water use by human activities will squeeze water out from the ecological system, reducing groundwater volume and causing a decline in groundwater table levels. The gap between water supply and water demand for ecological systems and the drop of groundwater table levels causes degradation and desertification of the woodlands and grassland, since there is a clear quantitative relationship between desertification, groundwater table and vegetation coverage according to existing scientific evidence (Liu 2000). Land degradation eventually constrains population growth and economic expansion which forms a complete feedback cycle. Moreover, population growth in the source region subsystem and the upper and middle streams' subsystems changes the water allocation and affects water sources available for the downstream subsystem.

Due to data limitation, the current system dynamic analysis only covers the period of 1980–1998. Using a system dynamics model, the historical changes in population, water sources, cultivated land and land degradation are simulated. A comparison of the actual values and the simulated results indicates a relative error in the range of -1.5 to two percent which is acceptable.

Using this model, we are able to disaggregate the effects of different factors on land degradation, and quantify the net effects due to population growth. The accounting strategy is simple: to calculate the changes in land degradation of a main subsystem under the scenario of no population growth, then compare it with the actual changes of land degradation so that the net effect of population growth is estimated. The general mathematical equation is as follows:

$$\Delta E_p/\Delta E = (\Delta E - \Delta E_0)/\Delta E$$

Where ΔE_p is the changes in land degradation due to population change, ΔE_0 refers to the actual changes in land degradation, and ΔE denotes changes in land degradation after excluding the effects of population growth.

Under the first scenario, we assumed that there was no population change in the whole Tarim River Basin over the past eighteen years, and obtained the consequences of land degradation in the downstream. Comparing with

Figure 9.13 Changes of land degradation under different population

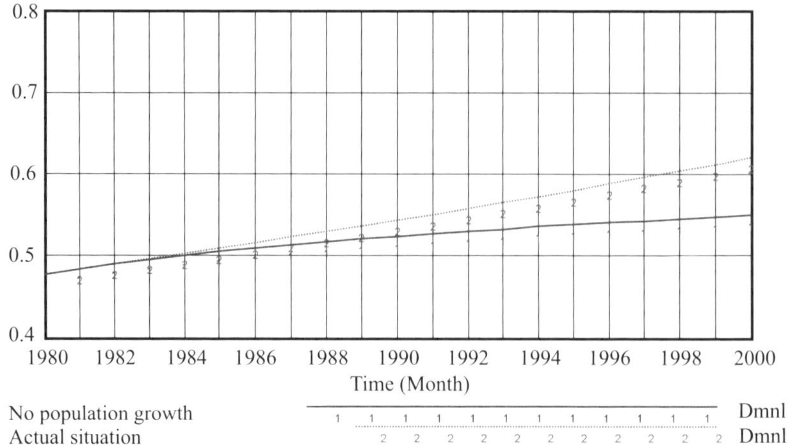

Figure 9.14 Net effect of population growth of different parts of the Tarim River Basin on land degradation in the downstream

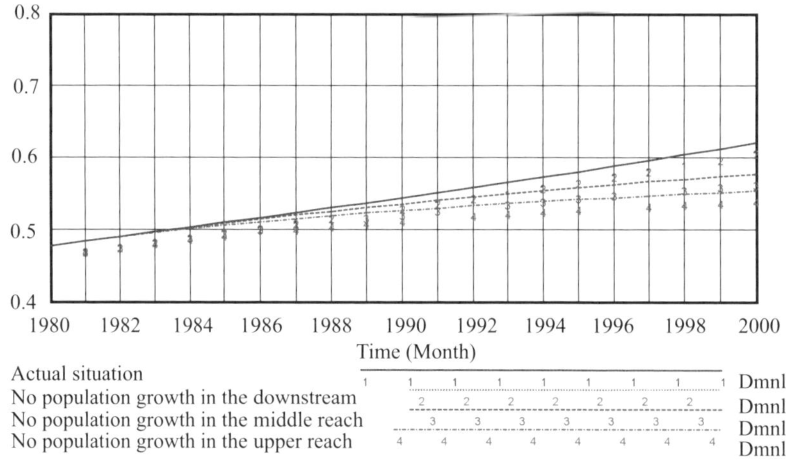

the actual statistics, our analysis indicates that population growth in the whole Tarim River Basin has contributed forty-nine percent to the land degradation in the downstream area in the past eighteen years (Figure 9.13). More specifically, population growth alone is responsible for 63.4% of grassland degradation, 25.9% of desertification and 13.1% of woodland deforestation.

To separately analyze the effects of population growth in different stream segments of the river, we take three other scenarios: 1) no population growth in the source region; 2) no population growth in the upper and middle reaches; and 3) no population growth in the downstream area. Simulation and comparative analysis shows that population growth in the source region is the main attributor and accounts for 46.5% of changes in land degradation in the downstream. Population growth in the upper and middle reaches also caused thirty percent of the land degradation in the downstream, while depopulation downstream eases the pressure of land degradation in that area by 0.14%.

It should be noted that the sum of the effects of population changes of all the stream segments on land degradation for the downstream does not equal the total effect of population growth of the whole Tarim River Basin. This is because population change only affects the land through other determinants, while the effects of those other determinants may cancel each other out. For instance, population growth in the source region reduces the water source for the upper and middle reaches which will hinder its land reclamation effort, while population growth in the upper and middle reaches itself induces the need for land reclamation. Therefore, the effects of population growth in the source region on the one hand and the effects of population growth in the upper and middle reaches on the other hand cancel each other out, and jointly affect the water availability and land degradation in the downstream. This also provides evidence that the environmental impact of population growth is not straightforward, and can only happen with other factors in a complex system.

Livelihood and out-migration downstream

Land degradation and environmental deterioration worsened the downstream living conditions. Given that the Tarim River Basin is a windy place—the annual strong windy weather (with wind speed of >17.2 m/s) accounts for about thirty days—sand storms happen very frequently as the land is degraded. The frequent wind storms destroy crops, block transportation vessels and cause human health problems. An investigation into the construction corps reveals that the incidences of respiratory and eye diseases dramatically increased after a sand storm. Moreover, worsening water quality also induced adverse effects on human health. Water-born intestinal and stomach diseases are very common downstream. An investigation conducted in the schools of Corp 32 revealed that forty percent of primary and middle school students suffered endemic dental fluorosis due to very high fluoride levels in the drinking water.

Driven by the adverse effects of land degradation—reduction of economic reward—people of the construction corps are seeking to leave this area. Most of the old generation of in-migrants are willing to send their children outside or back to their hometown in the east. Young people receiving education in the cities rarely go back to Yuli. Consequently, the population of the construction corps is aging rapidly. Most of the companies are encountering labor shortage problems which also contribute to the abandonment of cultivated land. All the corps are trying to attract in-migrants and itinerants by offering good benefits. However, those policies seem not to be effective. Only a few in-migrants, who are less educated and unskilled or seek to evade the more strict family planning policy in their hometown, come to the area. After coming, most of the migrants find that it is hard to stay, and quickly move out. Those out-migrants often move to the upper reaches of the Tarim River Basin. Recognizing the decline of population, the corps started to implement a family planning policy that allows young couples to have a second child. However, few couples want to have more than one child, facing the adverse environmental conditions. The fertility rate in the corps of Yuli County was even lower than that in many cities of the eastern provinces. Family planning officials complained that they changed their jobs from controlling births to encouraging births, with the latter task apparently being more difficult. Even those of the floating population who came to seek a chance of having more babies quickly changed their mind soon after arrival. With an aging population structure and increasing prevalence of diseases due to environmental deterioration, the crude death rate shows an upward trend. The increasing death rate and decreasing fertility rate, combined with negative net-migration, is shrinking the population size downstream. While many people are worrying about the continuation of the construction corps, during our interview, some officials of the construction troops mentioned that it would not be a bad idea to remove the construction troops and return the land to nature.

Conclusion

Historical inspection of the changes in population and land characteristics of the Tarim River Basin indicates that a large volume of in-migrants to the less populated arid areas caused serious land degradation problems. The intruding migrants drained the river stream, which was previously used as ecological water, to the irrigation system. As a result, the natural ecosystem has been transformed into a man-made environment. The landscape is commonly dominated by human settlements and desert areas, while the natural arable areas are being converted into either man-made or deserted environments. The ecosystem becomes much more vulnerable. The human population increasingly faces the challenges of pressing sand dunes. Losing the natural buffer zone, the human settlements were easily and directly turned into desert areas.

Compared with the indigenous groups, in-migrants are more 'exogenous' in terms of not only their origins, but also their production and consumption styles and the ability to adapt to changes in environmental conditions. Our study reveals that the environmental impacts of new in-migrants are more significant than that of indigenous populations. Conversely, we also found out that the new in-migrants are less able to endure the difficulties caused by environmental degradation. The shorter the residential duration, the more likely the migrants leave when facing environment changes.

In a dryland river basin like the Tarim, a general historical migration pattern might occur in the following manner: a trend of human settlement gradually and continuously migrates to the upper stream where water quantum and quality is always better. Consequently, the population density in the upper stream gradually increases, while water reaching the downstream areas decreases and the river length is continuously shortened. Therefore, a study on the interactions between population migration and land degradation in the river basin of the dryland should consider the dynamics in the whole river system. Population migration in the upper reaches may not induce land degradation in situ but may cause land degradation downstream; similarly, land degradation in the downstream reaches may have nothing to do with population growth due to migration in situ, but may rather be caused by population growth in the upper stream. Water as the key resource forms the linkage between population migration and land degradation crossing administrative boundaries. In stipulating an environmental policy against land degradation, water management regulations which systematically tackle the whole river basin are needed.

Land degradation in the Tarim Basin was caused by excessive in-migration and land reclamation. However, these two factors are themselves caused by faulty policy and market failure. The past experiences of the Tarim River Basin show that when stipulating and implementing economic development policy decision-makers should pay close attention to plausible environmental consequences. In particular, as China is implementing the West Development Strategy and establishing a market-oriented economy, it should be aware that the market cannot be fully responsible for environment protection. Moreover, embracing the new economic opportunities and a warmer and wetter climate phase since the end of the 1990s, the Tarim River Basin may face a new wave of unchecked in-migration and excessive land reclamation. There are probably many good reasons to adopt the idea of some local people who are suffering from the land degradation: 'it might not be a bad idea to return the land to nature'.

References

Bates, D. C. (2002), 'Environmental refugees? Classifying human migrations caused by environmental changes', *Population and Environment*, 23 (5).

Bilsborrow, R. (2002), *Migration, population change and the rural environment*, ECSP Report Issue 8, summer.

Bilsborrow R., and P. Delargy (1991), 'Land use, migration and natural resource deterioration: The experience of Guatemala and the Sudan', *The Population and Development Review*, 16: 125–147.

Darkoh, M.B.K. (1982), 'Population expansion and desertification in Tanzania', *Desertification Control*, 6: 26–33.

Deng, L. A. (2000), *Integrating Drylands Issues into Country Assistance Strategies: A Review*, The International Bank for Reconstruction and Development/The World Bank.

Ericson, J., M. Freudenberger and E. Boege (1999), *Population Dynamics, Migration, and the Future of the Calakmul Biosphere Reserve*, Washington, DC: American Association for the Advancement of Science.

FAO (1997), *State of the World's Forests*, Oxford: Woods and Publications.

FAO (2001), *The Global Forest Resources Assessment 2000: Summary Report*, Rome: FAO, Committee on Forestry *www.fao.org/forestry/fo/fra.*

Findlay, A. M. (1996), *Population and Environment in Arid Regions*, Policy and Research Paper No. 10, IUSSP Scientific Committee on Population and Environment.

Geist, H. and E. Lambin (2001), *What Drives Tropical Deforestation? A Meta-analysis of Proximate and Underlying Causes of Deforestation Based on Subnational Case Study Evidence*, LUCC Report Series No. 4.

Hecht, S. (1985), 'Environment, development, and Politics: Capital accumulation and the livestock sector in Eastern Amazonia', *World Development*, 13 (6): 663–684.

Hecht, S. and A. Cockburn (1990), *The Fate of the Forest*, New York: Harper Collins.

Heckandon, M. S. and A. Mckay (eds) (1984), *Colonization and Destruction of Forests in Panama: Essays on serious ecological problems*, Panama Assoication de Antropoligia.

Heyzer, N. (1995), 'Gender, population and environment in the context of deforestation: A Malaysia case study', *IDS Bulletin*, 26 (1): 40–46.

Hu, H. (1935), 'Distribution of China's population', *Geographical Journal*, 2.

Hugo, G. (1996), 'Environmental concerns and international migration', *International Migration Review, Special Issue: Ethics, Migration, and Global Stewardship*, 30 (1).

Ibrahim, F. (1987), 'Ecological and land use changes in the semiarid zone of the Sudan', in P. Little and M. Horowitz (eds), *Land at Risk in the Third World*, Boulder: Westview Press.

Ji, F. (2001), *Salt-Water Dynamics of Oasis Soil and its Management in Tarim Basin*, Beijing: Ocean Press

Joly, L. G. (1989), 'The conversion of rain forests to pastures in Panama', in D. Schuman and W. Partridge (eds), *The Human Ecology of Tropical Land Settlement in Latin America*, Boulder: Westview Press.

Knerr, B. (2001), 'Interactions between desertification and population development', working paper, Deptartment of Development Economics and Agricultural Policy, University of Kassel.

Leighton, M. (1999), 'Environmental degradation and Migration', in *Drylands, Poverty, and Development*, proceedings of the June 15 and 16, 1999 World Bank Round Table.

Li, P. (2000), 'Groundwater, vegetation and land desertification', in Y. Song, Z. Fan, Z. Lei and F. Zhang (eds), *Research on Water Resources and Ecology of Tarim River*, Xinjiang People's Press: 241–249.

Little, P. (1987), 'Land use conflicts in the agricultural/pastoral borderlands: The case of Kenya', in P. Little and M. Horowitz (eds) *Land at Risk in the Third World*, Boulder: Westview Press

Lonergan, S. (1998), 'The role of Environmental degradation in population displacement', Research Report 1, Victoria: University of Victoria.

Mahar, D. J. (1979), *Frontier Development Policy in Brazil: A Study of Amazonia*, New York: Praeger.

Moran, E. F. (1981), *Developing the Amazon*, Bloomington: Indiana University Press

Oberai, A.S. (1998), *Land Settlement Policies and Population Redistribution in Developing Countries: Achievements, Problems, and Prospects*, New York: Praeger.

Pfaff, A. S. P. (1997), 'What drives deforestation in the Brazilian Amazon?—Evidence from satellite and socioeconomic data', Policy Research Working Paper, The World Bank.

Pichon, F. (1997), 'Colonist land allocation decisions, land use and deforestation in the Ecuadorian Amazon frontier', *Economic Development and Cultural Change*, 45 (4): 707–744.

Pichon, F. and R. E. Bilsborrow (1999), 'Land use system, deforestation and demographic factors in the humid tropics: Farm-level evidence from Ecuador', in R. Bilsborrow and D. Hogan (eds), *Population and Deforestation in the Humid Tropics*, Liege: International Union for the Scientific Study of Population.

Schelhas, J. (1996), 'Land use choice and change: intensification and diversification in the lowland tropics of Cost Rica', *Human Organization*, 15 (3): 298–306.

Schmink, M. and C. Wood (1993), *Contested Frontiers*, New York: Columbia University Press.

Song, Y., Z. Fan, Z. Lei and F. Zhang (eds) (2000), *Research on Water Resources and Ecology of Tarim River, China*, Xinjiang People's Press.

United Nations Development Program (UNDP) (1999), Human development Report, *www.undp/hdro/population.htm.*

Wood, C. H. and D. Skole (1998), 'Linking satellite, census, and survey data to study deforestation in the Brazilian Amazon', in D. Liverman, E. Moran, R. Rindfuss, and P. Stern (eds), *People and Pixels*, Washington D. C.: National Academy Press.

World Wildlife Fund (2001), 'Tarim Basin deciduous forests and steppe', *http://www.worldwildlife.org/wildworld/profiles/terrestrial/pa/pa0442_full.html*

World Bank (1999) *World Development Report*, Washington.

Zheng, Y., Y. Tong and X. Yuan (1996), *Population Migration Study of Xinjiang*, Xinjiang People's Publishing House.

Zhou, H., Y. Song and S. Hu (2000), 'Irrigated agriculture and sustainable water management strategies in the Tarim Basin', in the proceedings of the workshop *New Approaches to Water Management in Central Asia*, held in Aleppo, Syria, November 6–11.

10
Water Shortage in an Oasis Region in Western China[1]

Nakawo Masayoshi

Water, the most valuable substance and the principal constituent of all living organisms, is a finite and vulnerable resource. Global water circulation has attracted much attention, and various research efforts are being devoted to the relationship between water resources and recent global changes. However, in arid and semi-arid regions, there is limited information available about the water circulation system where human activities are taken into account. In these extremely dry places, human survival is strongly dependent on a very limited amount of water.

Central Eurasia consists of arid and semi-arid regions, where people dependent on oasis farming or nomadic lifestyles have been active historically. This region has been a conduit for knowledge and goods representing the cultures of the West and the East, furthering and forging intercultural exchange. Sugiyama (1997) has pointed out that it is not only a transient region, but one of outstanding cultural forms as local people successfully assimilated knowledge and developed historically influential cultures, such as those of the Huns, Turks and Mongols.

Recently, water shortage has plagued western China, forcing the local people to change their traditional lifestyles. This shortage could be attributed either to recent climate change or changes in the social system, but a quantitative analysis of the situation is yet to be undertaken and the exact cause or causes are yet to be disclosed.

Water availability and global change

Figure 10.1 schematically represents the present river systems in the Tarim Basin with the distribution of present and past cities. It is found that most cities today are located on the piedmont area of the Tianshan and the Kunlun Mountains at sites adjacent to major rivers, facilitating utilization of the river water. Most local people depend on the river water for survival, because the precipitation in these cities is less than several tens of millimeters per year.

Figure 10.1 River systems in the Tarim Basin with present and past cities

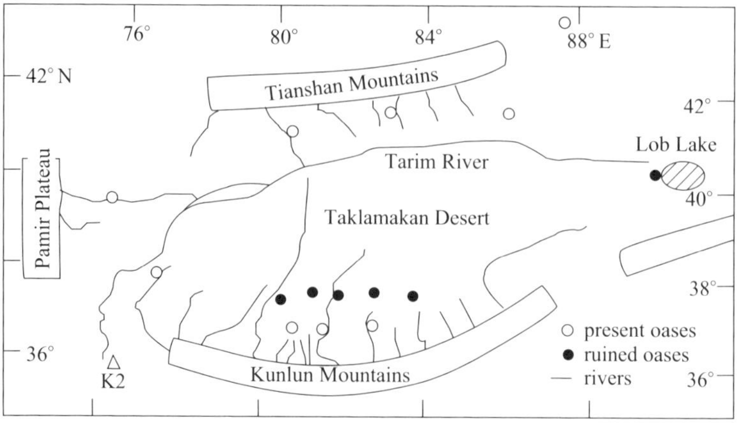

In contrast, the ancient ruined cities are further from the mountains and most are no longer next to rivers. Several cities, therefore, seem to have been relocated upstream historically in the southern margin of the desert. This would indicate that the flow-rate of rivers from the Kunlun Mountains has decreased, so that the people, and hence cities, moved upstream to obtain sufficient water for living, since less water became available at the original sites.

By examining old documents, Hoyanagi (1966) reconstructed historical variations of river flows on the southern side of the basin. He compared his results with the temporal change of sea level as shown in Figure 10.2. The flow-rate was found to have a negative correlation with the change in sea level: the amount of flow-rate is small when the sea level is high, and vice versa. Sea level and global temperature are considered positively correlated, as recent global warming has accelerated the melting of land ice mass, and accordingly increased the sea level over the last 100 years (IPCC 2001).

In the arid and semi-arid regions of central Eurasia, glaciers supply a significant amount of water for most of the rivers. Ujihashi and Kodera (2000) showed, for example, that water from glaciers accounts for more than fifty percent, on average, of the total flow-rate of the Yurungkax and Keria Rivers originating from the Kunlun Mountains. The amount of river flow-rate in the region should therefore be strongly affected by the change in air temperature. This seems to contradict the correlations discussed in the previous paragraph, if we try to explain the change in flow only in terms of glacier melt.

A change in precipitation per se should also cause a change in availability of water from mountains, which is the major water resource for rather dry areas

Figure 10.2 Change in river discharge in the southern Tarim Basin versus sea level change

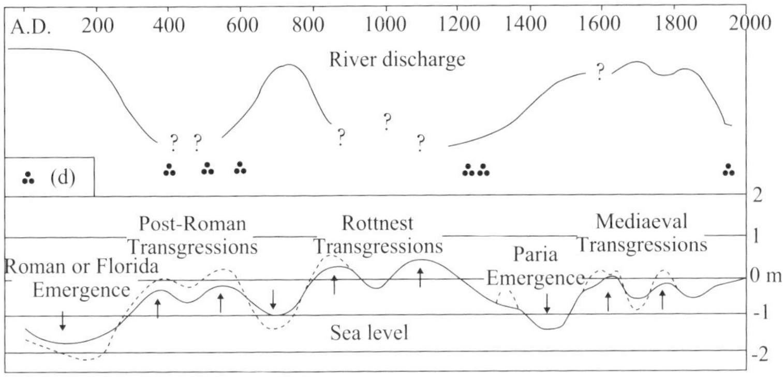

Source: Hoyanagi (1966).

at low elevations with little precipitation. It has been suggested that the global average precipitation would increase as a result of global warming (IPCC 2001). The spatial distribution of precipitation in any given region, however, is complex and highly dependent on the local and surrounding geographical conditions, and an increase of the globally averaged precipitation does not mean an increase in precipitation for all regions. In fact, there are several studies that report a decrease in precipitation in association with an increase in temperature, particularly in arid and semi-arid regions (e.g. Nakawo et al. 1990). Nevertheless, precipitation change is strongly correlated with the global water circulation system.

Water Shortage in the Heihe Basin

The Heihe River, the second largest inner river in China, flows from a glaciated zone in the Qilian Mountains north through Gansu Province and disappears in terminal lakes in Inner Mongolia. The total length of the river is about 400 km with a basin area of approximately 128,000 km^2 inhabited by about 1.8 million people (Pan and Tian 2001). The basin is divided into three zones depending upon altitude. The upper reaches are a mountain zone above 2500 m above sea level with annual precipitation of more than 200 mm. Here, Yugur people live, depending mainly on stock farming. The middle reaches are 1500 m to 2500 m in elevation with an annual precipitation of about 100 mm. The area includes several oases, where many Han people have migrated to, depending primarily on irrigation agriculture (oases zone). The lower reaches are a desert zone with annual precipitation of less than fifty mm. Here Mongol people live, reliant on stock farming for their livelihoods.

It was recently found that terminal lakes of the Heihe River have disappeared (Wang and Cheng 1999). Also, the vegetation in the lower reaches has become increasingly poor, and many shallow wells have dried up. Thus, the water shortage has become obvious, especially in the lower reaches. Some local informants claimed that the shortage is caused by over-grazing of livestock kept in the lower reaches inhabited by Mongolians. They often blame over-grazing by livestock kept by Yugur people as a cause of the deforestation in the mountains that they say decreases the water supply flowing into the lower lands. To preserve the forests, the Chinese government adopted a policy called 'environmental relocation', that requires nomadic people in the mountains to move to an urban area.

Figure 10.3 shows changes in the river flow-rate for the last half a century at two gauging stations: one at the foot of the mountain between the upper and middle reaches, and the other at the exit point from the oases zone to the desert zone located between the middle and lower reaches. It can be seen that the amount of river flow into the oases zone from the mountain zone has increased slightly, while outflow from the oases zone to the desert zone has decreased significantly over the same time period. This indicates that water use in the middle reaches (oases zone) has increased very rapidly. The difference in flow-rate at the two stations is plotted in Figure 10.4 in order to show the water consumption in the oases zone during the period from 1955 to the late 1970s.

The amount of water used in the middle reaches was about five m^3/s (around 0.16 Gt per year) during the period from 1955 to the late 1970s. Then it increased very rapidly, doubling from 1980 to 2000 when annual water use was about ten m^3/s (around 0.32 Gt per year).

This increase seems to synchronize with population growth in the basin, and be caused by convenient manipulations of available water: dam construction and the development of irrigation systems mainly for agriculture. The current population is about 1.8 million, of which 1.4 million people are farming mostly in the middle reaches where water is relatively easy to obtain.

Agriculture depends not only on river water, but also on groundwater. The intake sites from the river are relatively low for high terraces of oases, with a height difference of about fifty meters in the case of Zhangye, one of the major oases in the Heihe Basin. At the higher area water is taken from wells for irrigation agriculture, except at small villages at the outlet of small tributaries including those found near Matisi, a small village located around

Figure 10.3 Change in rate of river discharge (unit: m^3/s)

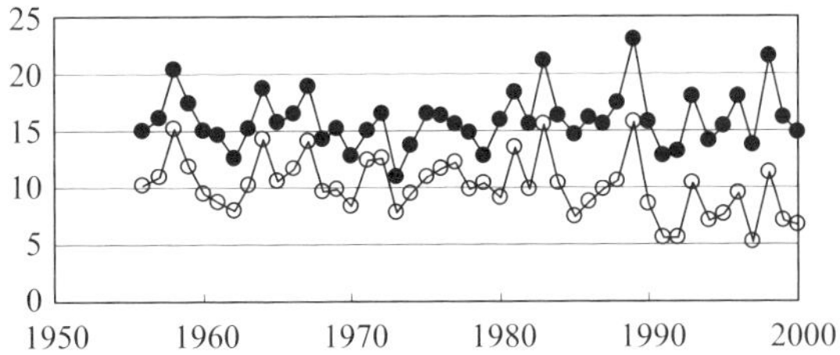

Figure 10.4 Estimated water consumption in the oases zone

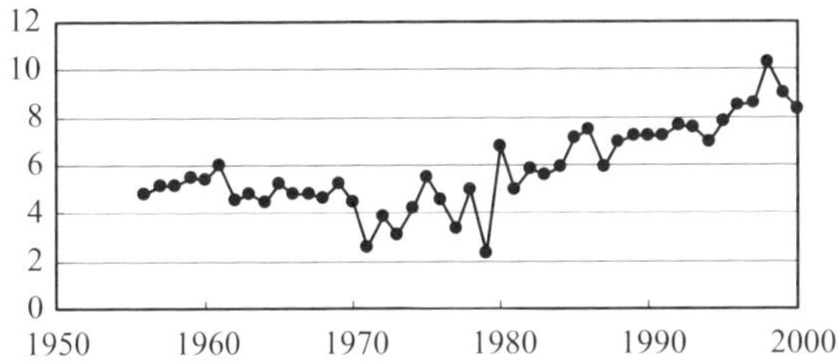

Note: Includes water loss by evapo-transpiration and groundwater recharge, based on the difference in water discharge data in Figure 10.3 (unit: m^3/s)

a Tibetan temple. As a result of groundwater use, the water level has recently decreased to varying degrees at several wells even in the middle reaches, as shown in Figure 10.5. Thus, the water shortage has become evident not only in the lower reaches, but also in the middle reaches.

In the lower reaches, where water scarcity is most evident, the annual precipitation is generally decreasing, as shown in Figure 10.6, although the inter-annual fluctuation is rather large. The gradual decrease might be one of the results of recent global warming.

As we have described above, the present shortage of water and deterioration of vegetation, in particular in the lower reaches of the Heihe Basin, could have been primarily caused by agricultural development in the middle reaches. Over-grazing both in the mountains and in the lower reaches could be other potential causes for recent deterioration of vegetation in the lower reaches. In addition to these human impacts, climate change could affect the availability of water in the region through global warming, a decrease in precipitation, an increase in evapo-transpiration and glacial shrinkage. These changes are interlinked, and their dynamic inter-relations plus the human impacts should be studied quantitatively to determine the most effective measures to alleviate the present water shortage.

Figure 10.5 Recent changes in groundwater level at selected wells in the oases zone

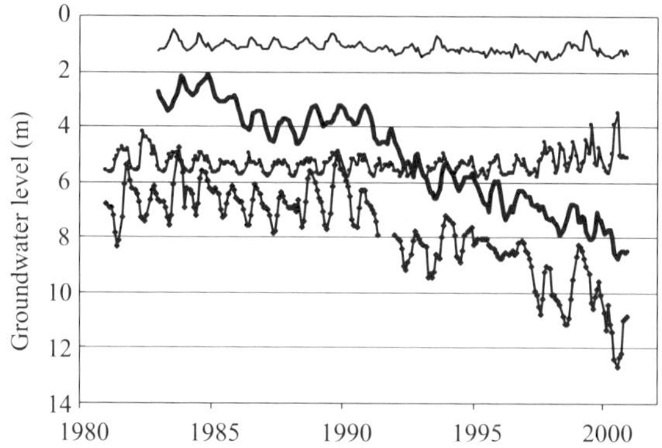

Figure 10.6 Change in annual precipitation at Ejina, central city in the lower reaches

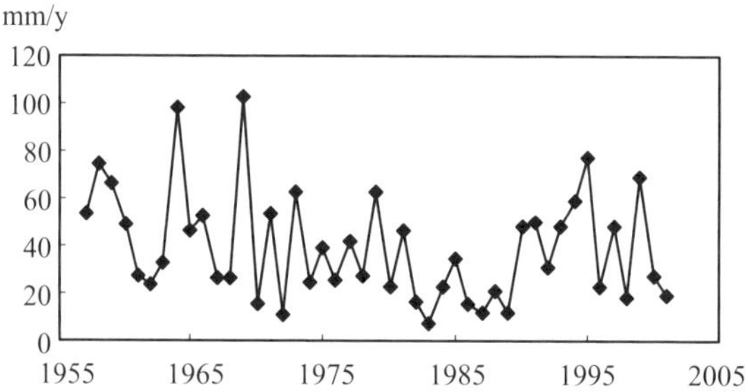

Historical view

There are many ruined old cities scattered around the basin, in particular in the middle and lower reaches. Near Zhangye city which used to be called Ganzhou, for example, there is a ruined city called Heishuiguo. Its southern city is thought to be founded in the era of the Northern and Southern Dynasties including the Wei and Jin Dynasties in the fifth to sixth centuries, due to the unearthing of a series of tombs from that era. The northern city of Heishuiguo is considered to have been established in the Han dynasty in the 2nd century BCE, older than the southern one. Luotuocheng ruin in Gaotai County is another example. A former oasis is thought to have extended around Luotuocheng, but only a desert and abandoned farmland are seen at present. Extensive ruins of old cities and tombs mostly of the Han and Tang Dynasties, such as Luotuocheng and Xushanwan, indicate intensive human activities there between the second century BCE and the tenth century AD.

The largest tributary of the Heihe River is Beidahe River (or Tuolaihe River). This river does not fully run through Jiuquan city, an old city that used be called Suzhou, since the river water is entirely extracted at a weir near Jiayuguan (constructed in 1964 and in operation in 1966). The director of the weir told us that the flow-rate had decreased every year. The Beidahe River used to join with the Heihe River (Tan 1982), but no longer does. This indicates that the volume of river water has decreased significantly, although the decrease could be largely due to water consumption at a dam near Jinta constructed in the late 1940s.

In the lower reaches, there are several old cities and old fire-signal towers, constructed in the Han dynasty (202 BC to 220 AD) for protection against invasion by the Hunnu. The Heicheng city in this area is considered to have been founded in the Xixia dynasty, and then enlarged in the Yuan dynasty. At Lucheng city, there are relics of old irrigation channels, evidence of the development of large scale hydraulic constructions. The distribution of many ruined cities and farmland, extending from the irrigation channels, indicates that the lower reaches of the basin were prosperous and had sufficient water at least at certain points in the history. Even in the beginning of the twentieth century, when many explorers visited the lower reaches, a very dense forest was present, which was recorded in a film taken in 1927 by a German cameraman who accompanied Sven Hedin's expedition.

Figure 10.7 shows an approximate population change in the basin from the Han to Ming Dynasties (Nakawo 2002). It can be seen that as many as about 1.5 million people moved into the basin intermittently in the Han, Sui-Tang and Xixia-Yuan Dynasties. Additional immigration involving large numbers of people took place in the Qing dynasty from the sixteenth to twentieth centuries. In other words, large groups of people have repeatedly migrated in and out of this basin region over a period of roughly 600 years (three peaks in Figure 10.7). Successive waves of people arrived in the Qing dynasty (fourth peak) and at present (fifth peak), since the current population is about 1.8 million, almost equivalent with that of the old days.

Some may say that a long term periodicity of population could be related to a global climactic influence, since there have been a few peaks in the global average temperature that might coincide somewhat with the peaks in population shown in Figure 10.7. For example, there is the Medieval Climatic Optimum near 1200 AD, another peak about 100 AD and the present relatively high global temperature. We do not know, however, that these warm epochs took place in every region of the globe. Also, some population peaks do not correspond with

Figure 10.7 Population change in and around the Heihe basin from the Han to Ming Dynasties

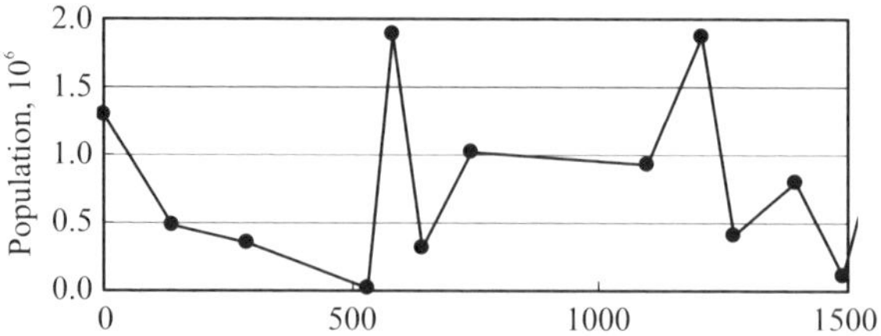

Source: Nakawo (2002).

Note: The data are from Gansusheng Dang'anguan (1977), and do not exactly represent numbers in the basin, but could indicate a rough change of the population in the basin.

a so-called warm period. People would certainly react and must have reacted to environmental changes. However, they are proactive in the environment as well. History must not and should not be determined by environment alone. Careful studies with wide spectrum and perspective are crucial and required.

Nonetheless, as past societies in the area used irrigation systems for agriculture, similar water shortages could have occurred in the Tarim Basin in the past while the population was as large as the present population. To obtain more clues to help solve the present water shortage problems it is hence important to examine the historical evolution of the inter-relation between natural and social systems in the basin.

Notes

1. This is a contribution from the Oasis-project (Historical evolution of adaptability in an oasis region to water resource changes) promoted by the Research Institute for Humanity and Nature. The manuscript was prepared in 2003.

References

Gansu Sheng Dang'anguan (1997), *Gansu Lishi Renkou Ziliao Huibian* (Historical population in Gansu Province), Lanzhou: Gansu Renmin Chubanshe.

Hoyanagi, M. (1966), 'Geographical problems concerning the Old Silk Road region in the Tarim Basin', *Geographical Report of Tokyo Metropolitan University*, 1: 1–32.

Inter-Governmental Panel on Climate Change (IPCC) (2001), *Climate Change—The Scientific Basis*, New York: Cambridge University Press.

Nakawo, M., Y. Ageta and J. Han (1990), 'Climatic information from the Chongce Ice Cap, West Kunlun, China', *Annals of Glaciology*, 14: 205–207.

Nakawo, M. (2002), 'Kokugaryuuiki shuuhen'no jinko hensen' (Population change in and around the Heihe Basin), *Project Report on an Oasis-region*, 2 (2): 183–185.

Pan, Q. and Tian, L. (2001) *Heihe Liuyu Shuiziyuan* (Water Resources in the Heihe Basin), Zhengzhou: Huanghe Shuili Chubanshe

Sugiyama, M. (1997), *Yuubokumin Kara Mita Sekaishi* (World history from the standpoint of the nomads), Tokyo: Nihon Keizai Shinbunsha.

Tan, Q. (1982), *Zhongguo Lishi Dituji* (Historical Atlas of China), Beijing: Zhongguo Ditu Chubanshe.

Ujihashi, Y. and S. Kodera (2000), 'Runoff analysis of rivers with glaciers in the arid region of Xinjiang, China', *Research Report of IHAS–Water in Arid Terrain Research*, 8: 63–78.

Wang, G. and G. Cheng (1999), 'Water resource development and its influence on the environment in arid areas of China: The case of the Heihe River basin', *Journal of Arid Environments*, 43: 121–131.

11
Marginal Capital: Beijing's Maturing Water Economy and Agricultural Sustainability

James E. Nickum

Introduction

Like New Delhi, the capital of the world's only other billion-body state, Beijing lives on a number of margins, prominent among them the hydrological one. Its eight urban districts, six suburban and periurban districts, and two districts and two counties in rural areas cover an extensive area of 16,800 km^2 at the northwest edge of the arid North China Plain. Beijing has over two-thirds the population of the Kinki Region (fifteen million cf. twenty-one million) residing in an area about half its size. Over three-quarters of its people live in an urban core that is virtually the same in extent as Kyoto Prefecture in Japan, but the terrain is much flatter than Kyoto and sprawls rapidly into periurban areas.

Beijing is the most favorably located city in north China in terms of water resources, with an average annual precipitation of 625 mm and the most abundant natural groundwater supply on the north China plain. Despite this apparent abundance, water has long been a major concern of governments of Beijing and its predecessor cities. One reason for this is that the average rainfall is quite limited in comparison with the more humid areas of southeastern China or Japan. For example, Kyoto Prefecture's annual precipitation is 2.5 times that of Beijing. In addition, precipitation varies widely from year to year, ranging from as little as 242 mm (recorded in 1869) to as much as 1506 mm (in 1959), with dry years in particular often coming in clusters. The Mongols, who established the capital Dadu of their Yuan dynasty (1271–1368) on the present site of Beijing, were concerned about water supply, flood prevention and using the Grand Canal to bring in grain from richer agricultural areas in the south. Their water plans have often provided the basis even for post-1949 works. Nonetheless, while the Mongols provided the urban area with water and some of its most scenic areas, traditional diversion works did little for the rural areas. Until recent memory, Beijing's agriculture relied on rainfall. Yields were unstable, famines occurred and water crises were a part of life.

The maturing water economy

In the 1950s and 1960s, it seemed that the situation had turned around, due primarily to the construction of major water storage facilities. Soon thereafter, however, water crises reentered Beijing's history on an almost perennial basis. To help understand the factors behind this transformation, it may be useful to introduce the simple heuristic of the maturing water economy.

The histories of urban water supply development can be roughly divided into two kinds of phases—expansion and maturation. An expansion phase commonly follows the addition of a major new water supply source, such as through construction of a reservoir or a river diversion. Because of the large fixed scale of these works in the expansion phase, water supply is elastic—it costs relatively little to get more. Policies tend to promote water use. For example, subsidies allow the provision of water at low or no cost to end users, usually farmers, and do not unduly deprive other, more valuable uses. There is little competition between different water resource uses, such as agricultural, industrial, domestic, municipal or environmental. The opportunity cost of water is low. Conflicts tend to occur over land use rather than water. New water facilities need little maintenance, although tight budgets, overstaffed supply organizations, and low cost recovery may lead to 'deferred maintenance' down the track. The environmental problem of greatest concern is often flooding and drainage.

In the mature phase, water supply is inelastic and increasingly costly. Subsidies for water use begin to evidence high social costs, as they privilege established users (usually relatively prosperous farmers) over competitive newer ones such as industry, who can often put the water to better economic use, or, ironically, those whose subsistence needs are greater, as in the case of many migrants to the cities. Aging infrastructure demands higher maintenance or replacement and a growing share of infrastructure funds is geared towards relieving stress on the system. Typical environmental problems are water pollution and the depletion of groundwater aquifers, both of which threaten the existing water supply.

Concomitant with this framework, the history of urban water construction may be seen as the result of a combination of: 1) struggles among stakeholders (social conditions); 2) the impacts of economic demands (economic conditions); and 3) dominant institutional arrangements (responses). Institutional rigidities such as vested interests often resist change until a point of crisis is encountered, such as a major drought or pollution event. As new sets of arrangements are shaped, the solutions to problems more often than not bring new sets of institutions and new sets of problems.

Water economies can rejuvenate by finding and developing new sources. The maturing water economy framework is thus not strictly unidirectional. New technologies can expand supply, as can major projects such as reservoirs, diversions or recycled water treatment facilities.

For example, in much of Asia, especially north China and the Punjab, the spread of deep pump tube wells in the 1960s and 1970s allowed an unprecedented tapping of vast aquifers. Areas that formerly relied on the whims of weather to eke out low crop yields became irrigated grain belts. In the cities, industries dug wells to provide for their own needs without having to rely on an outside supplier, such as the government. It should be noted, however, that water from tube wells does entail sometimes significant incremental costs in fuel, electricity and maintenance.

Bejing's expansion phase

We can locate an expansion phase in Beijing's modern history with the building of its two large reservoirs, the Guanting in 1954 and the Miyun in 1959. Ironically, floods were at least as much of a concern as drought in the construction of these impoundments. Waters from these reservoirs were so abundant that they also supplied Beijing's downstream neighbors of Hebei Province and Tianjin Municipality.

A major flood in 1963 in the Hai River basin, where Beijing is located, followed by a wet 1964, only reinforced the view that excess water was a more serious threat than water shortage. The municipality's economic planners did not consider water a limiting factor in converting the capital from a 'non-productive' administrative city to a 'productive' industrial center with a relatively complete set of modern, water-using enterprises that also required a lot of thermal power, itself a major water user given the low level of reuse at the time. The largest industries relied on their own wells for most of their water supply. There seemed to be enough water to accommodate urban growth. Beijing's urban water delivery capacity increased significantly in the first forty-five years of the People's Republic.[1] Then came irrigation, followed by the dry years.

The end of expansion: Signs of stress

The stresses on urban Beijing's water supply became apparent in the 1970s when the water table fell rapidly and the artesian sources in the western hills stopped flowing. By 1980, the aquifer was pumped down to bedrock in one location (Figure 11.1). Elsewhere, the water table continued to recede at an average of half a meter per year. Equally ominously, water flowing into the two large reservoirs went into decline because of climate change and, in the case of Guanting, due to increased upstream use and siltation. Rural areas in particular cut back their surface water use drastically in the 1980s.

Financial resources also began drying up with the economic reforms that began to take effect in that decade. Water had been supplied at heavily subsidized prices, with government subsidies providing the principal source of revenue to delivery agencies. Then funds to the Beijing Municipal Water Bureau were cut below the level necessary for renovating the water system, and one-fifth of its management personnel were released (Hu and Nie cited in Nickum 1997: 45).

Beijing's further growth should have been severely limited by its water conditions, yet the capital's economy, like that of the rest of China, has exploded over the past two decades (Figure 11.2). Even agriculture has expanded and diversified out of grain and a dwindling number of farmers have prospered (Figures 11.3 and 11.4). How did this come about?

Figure 11.1 Beijing's falling water table

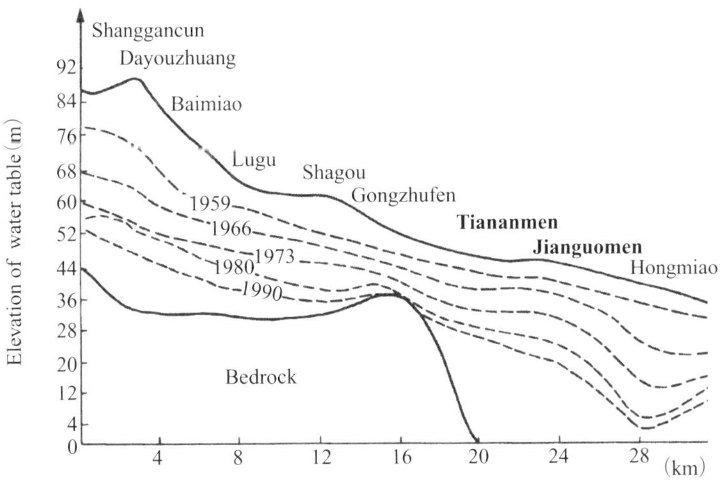

Source: Beijing Municipal Water Resources Bureau (1999: 14).

Figure 11.2 Beijing GDP

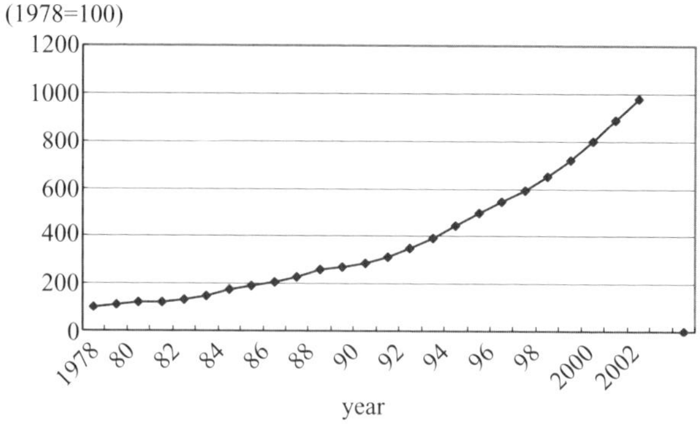

Source: Beijing Municipal Statistical Bureau (2004: 85).

Figure 11.3 Beijing irrigation and grain

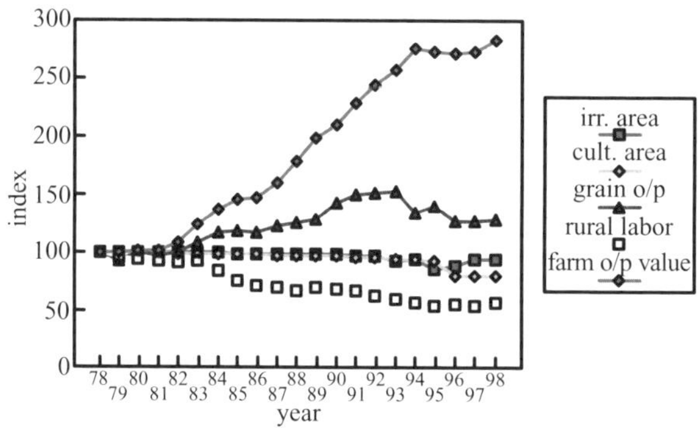

Sources: *Zhongguo shuili nianjian* (Yearbook of China Water Resources) (2003: 448); China Agriculture Statistical Report (2003: 18).

The transition to a maturing water economy

In response to the water stresses, Beijing's authorities (and those of the central government) sought, often on an *ad hoc* basis, both to follow the traditional path of increasing supply and, more importantly, to take measures to control demand. Among the latter measures were reallocating water supplies, cutting off much of the remaining supply to rural areas, reducing water use in urban areas and beginning efforts towards more integrated basin-wide water resource planning.

Figure 11.4 Income and output trends

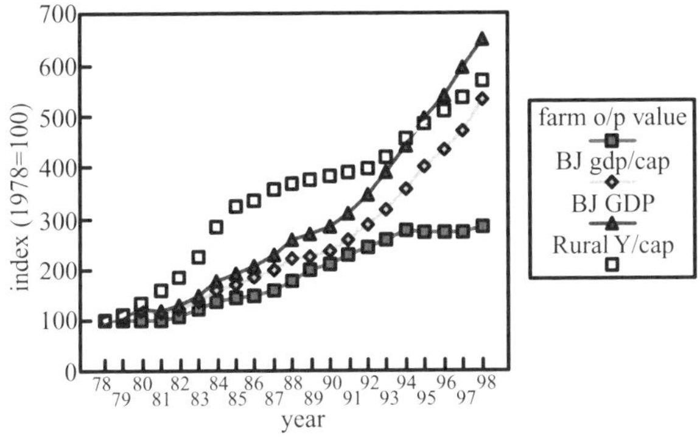

Sources: Li Yanchun, 'Gaigekaifang yilai GDP fanfan fenxi' (Analysis of doubling of GDP since the reform and opening up) in *Beijing Shi Shehui Jingji Tongji Baogao* (2004: 85).

Claiming upstream rights

Except in times of emergency, China does not appear to have any effective mechanism like the interstate compacts of the United States that allow users in lower reaches to establish and enforce prior rights if people upstream wish to increase their withdrawals. The readjustment of Beijing's boundaries in the late 1950s to include the Miyun Reservoir and thirty percent of its watershed no doubt made it easier for the capital to claim more exclusive rights over the water. Nonetheless, it has been necessary for the municipality to negotiate with the upstream province of Hebei over the quantity and quality of water flowing into the Miyun Reservoir. Even though these negotiations reportedly involve compensation payments, the results have been mixed in the absence of an established system of long-term water rights (Peisert and Sternfeld 2005: 37–38). Beijing was even less fortunate in protecting the inflow into the Guanting Reservoir, which lies almost entirely inside Hebei, from upstream diversions.

At the same time, efforts have often been made to compensate downstream areas with compensatory water, yet the timing appears to be more ad hoc than planned. In 1981, the State Council allocated the water of the Miyun Reservoir entirely to Beijing. A compensatory reservoir was built for downstream Tianjin, which faced an even more desperate water supply situation, but it had not yet been completed at the time of reallocation. In the interim, Tianjin was forced to rely on two expensive emergency diversions from the Yellow River to provide it with adequate water.

Cutting farm usage

With a reduction in surface supply from the two major reservoirs following the very dry year of 1980, Beijing's farmers turned to groundwater. In that one year, the amount of water they took from the aquifers increased from 983 million m^3 to 1691 million m^3, as rural surface water use collapsed from nearly 2000 million m^3 to about 750 million m^3—dramatically reversing the proportions drawn from the two sources. Even with this drastic switch in sources, the total amount of water available to farmers fell by nearly one-fifth. Subsequently, in 1984, the State Council authorized the further transfer of water from the Miyun Reservoir from irrigation to urban purposes. The diversion project allowing this to occur was not actually completed until six years later, giving farmers some time to convert to more water-saving irrigation techniques such as sprinklers. To encourage this conversion, the state and villages provided over ninety percent of the conversion costs, leaving farming households with only a small direct financial burden for setup costs.

Irrigation is primarily used to produce grain (maize and wheat, plus some rice), and secondarily to grow vegetables. Beijing's grain output increased from 2.18 million metric tons (mmt) in 1984 to 2.65 mmt in 1990. Since then, grain production has fallen, reaching 2.32 mmt in 1997 and collapsing to 580,000 tons in 2002 (Ministry of Agriculture 2004: 27).[2] One of the reasons for this decline is no doubt the reduction in water supply. A related factor is the effective increase in the cost of available water, due to higher pumping costs and changes to field crop irrigation using high-tech methods such as sprinklers.

It is important to note, however, that grain production has lost its significance as a source of employment or income to Beijing's rural residents. By 1997, only forty-one percent of a declining overall rural labor force of 1.6 million was primarily engaged in farming of any sort (including raising livestock), compared to sixty-eight percent of 1.9 million in 1984. In 1984, grain provided 14.5% of all income in Beijing's rural areas, while agriculture in total provided 24.1%. By 1996, all agriculture, including crops, forestry, livestock and fisheries was the source of a mere 7.3% of rural income (Figures 11.3 and 11.4).

Township and village enterprises (TVEs) and income sent by family members working in the city have come to dominate the employment and income scene in rural Beijing. TVEs place relatively modest demands on water resources when compared to irrigated agriculture, and add far more value per cubic meter. The downside is that there was a disturbing tendency to establish or relocate highly polluting industries such as paper mills, tanneries and chemical factories in the countryside. This has presented local people and officials with a difficult choice between employment opportunities and jeopardizing the groundwater they have come to rely upon.

Efforts have been made both to change the cropping structure and to reduce water use rates on existing crops. In 2000, the municipal government issued two regulations promoting more economical water use, especially in agriculture: the area devoted to growing rice and winter wheat was further reduced and farm tube wells were all equipped with meters.

Reducing urban use rates

The Beijing municipal government initiated a major campaign to reduce industrial water use rates in the early 1980s. First, it identified the 259 enterprises using over 100,000 m^3/year and targeted them in 1981 with 'planned water supply measures'. The following year the government introduced a steeply rising block-rate water fee system, indicating that those measures included monitoring the wells used by the enterprises. Use rates dropped markedly and sustainably. Beijing entered into a trajectory similar to that of Japan and the USA where total industrial water withdrawals have continuously declined over time, even as output has continued to increase.

In the meantime, water meters were installed at the urban household level rather than in apartment blocks as before. These allowed water fees to be raised from 1.76 *yuan* per cubic meter in 1999 to 4.07 *yuan* (in principle) in 2003. While these devices may have served to constrain the growth in domestic water use during a time of rising income and shifts toward more water-intensive lifestyles, their impact on overall demand has been limited by the relatively insignificant amount of water used by residences, even within the city proper.

The *Sustainable Use Plan* for Beijing's water for the period 2001–2005 (Beijing Shi Renmin Zhengfu 2001) called for a further increase in water fees, to an average of six *yuan* per m^3 for all users and 4.5 *yuan* per m^3 for domestic users. These included sewerage changes as well as water supply fees. A Beijing municipal official claimed that the city's water rates had reached 5.04 *yuan* per m^3 in 2004, the ninth rise since 1991 (Li Jing 2004). Yet a later source indicated that average water fees had only reached a 3.7 *yuan* per m^3 in 2004 (*Xinhua she* 23 December 2004), and there was considerable resistance to further increases. Even the target level would only be one to two percent of disposable income, probably not enough to make a major difference by itself on household water use behavior (Nickum and Lee 2006: 241).

Block pricing schemes, where the price of water rises if use exceeds a quota or graduated set of quotas, have long been accepted by policy makers as a means of limiting excessive use. A proposal to apply such a scheme to households in 2004 was postponed, however, because of metering and household registration problems (Li Jing 2004). Prepaid integrated circuit cards have become common for other utilities as a way of limiting use, but have not been adopted for water, also largely due to metering problems (Hu Tao, personal communication, 2005).

Rapid economic growth in a mature water economy

After 1984, despite continuing rapid economic growth, the water crisis in Beijing appeared to have abated somewhat. For one thing, precipitation increased (Figure 11.5). At the same time, as noted above, industry and agriculture both reduced their water usage without apparent detriment to growth in jobs and incomes. One positive outcome of the reduction in water use by industry and households was a relatively slow growth of less than one percent per year in the discharge of wastewater. Nonetheless, since most of this water is not treated before discharge, quality degradation has been at least of equal concern as conflict over quantity.

A decade ago, almost all of Beijing's wastewater was discharged without treatment. By 2003, more than half of the annual discharge of 1.2 billion m^3 was treated in a number of large water treatment plants. By the year 2008, over ninety percent of the city's sewage was to be treated by thirty plants.

Chinese analysts point to a number of abiding economic and institutional problems. Water fees are considered too low, especially for Beijing's industrial users. In addition, administrative fragmentation and fiscal rules contribute to a separation of the incidence of costs from projected benefits, inhibiting new investment, especially in conjunctive uses of surface and groundwater. Economic growth has also brought with it demands for recreation and watershed developments that require greater attention to watershed protection above the major reservoirs.

The latest enduring drought

Every year from 1999 to 2007, precipitation was below normal, creating a prolonged stress on an already stressed system. In particular, low runoff has severely cut surface water supplies beyond the reductions necessitated by declines in inflow quantity and quality from the upper reaches of the Miyun and especially the Guanting reservoirs (Figure 11.5). In addition, concerns

Figure 11.5 Water resource availability in Beijing

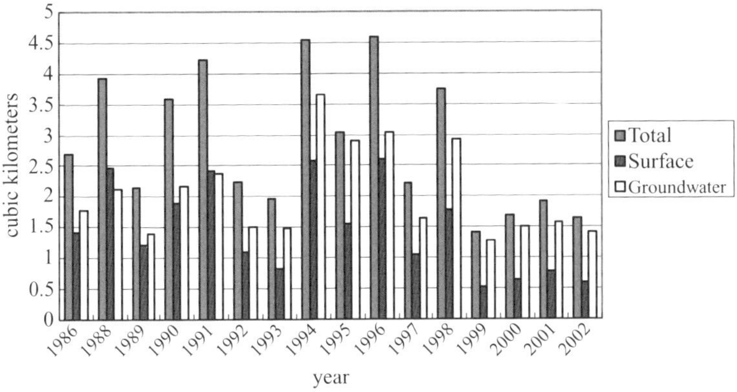

Sources: Lu Zhiguang et al. (2002: 4) (1986–99); *Beijing Water Resources Bulletin 2001* (2001: 2, 8, 11); *Yearbook of China Water Resources 2003* (2003: 300) (2002).

grew about Beijing's ability to provide adequate water for visitors to the 2008 Olympic Games. In addition to the pricing schemes mentioned above, China has addressed these problems with a multi-faceted approach.

Water-saving appliances

Valves and taps in public institutions and households in Beijing have been quite leaky in the past, and flush toilets have had nine liter or greater tanks. The widespread introduction of water-saving appliances such as improved valves and shower heads and six liter toilets contributed significantly to a reported reduction of municipal water use from 4.1 billion m^3 in 2001 to 3.46 billion m^3 in 2004 (Xinhua Net 2005).

Water-saving agriculture

Conventional irrigation is marked by high losses in transmission between source and crop root zone. As shown in Figure 11.6, nearly all irrigated land in Beijing has been equipped in some way to reduce these losses, although lining is controversial as it prevents seepage that replenishes groundwater. In addition, the structure of cropping has been altered drastically in recent years, with grain output plummeting and irrigation being diverted to higher-value vegetable crops or fodder for the booming livestock industry.

Authorities have been much more reluctant to increase the price of water for agriculture and farming households, even though they account for forty-five percent of Beijing's water use. There are a number of reasons for this, including metering problems, lower incomes in rural areas and the political

Figure 11.6 Water-saving irrigated area, 2002

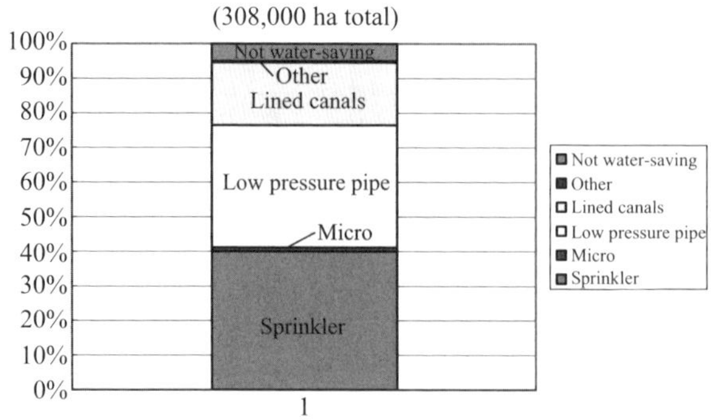

Source: *Yearbook of China Water Resources 2003* (2003: 450).

volatility of increasing the burden of official levies on farmers. Even in urban areas, as noted above, there are abiding problems with outdated meters and inaccurate household registration (Li Jing 2004).

Augmenting supply

Under a program adopted by the central government in 2001, large amounts of water are being transferred from nearby provinces, especially Hebei, beginning in 2003 and reaching 410 million m^3, or half the city's total tap water, in 2005 (Nickum and Lee 2006: 235). The central and municipal governments allocated a special fund of twenty-two billion *yuan* (US $2.7 billion) for these diversions and compensation to the areas affected in Shanxi and Hebei provinces (*China Daily* 2003). It was hoped that these would ensure adequate water for the 2008 Olympics, which, to the relief of all, was carried off successfully. After 2010, it is expected that Beijing will benefit from a large-scale 1,200 km diversion of up to 1.3 billion m^3 per annum of water from the Yangtze (Wang Hao et al. 2003: 147).

It should be noted that Beijing's economy has continued to boom during the drought years. The city has not entirely turned away from a supply augmentation strategy. There are indications of continuing serious over-pumping from deep periurban rural aquifers, where the battle between urban water uses may have literally moved underground.

Economic restructuring

In agriculture, cropping structure has moved sharply away from grains (especially wheat and rice) towards vegetables and fodder crops. Between 1995 and 1998 the percentage of sown area devoted to grains fell from 78.5

to 52.9%. Because feed crops (especially alfalfa) and vegetables are water-intensive, the net effect on total water demand was not as dramatic (Lu, Sun and Lu 2002: 111). There appears to have been a mild waver in this trend (see footnote two), with grain being pushed at the possible expense of vegetables and livestock, but this appears to go against the long-term trend.

Bureaucratic reform

In March 2004, the Beijing municipal government created a comprehensive Municipal Bureau of Water Affairs based on the previous Bureau of Water Resources. This new entity also assumed responsibility over water supply, water conservation, wastewater treatment and use and conservation of underground water that had previously been handled by the Municipal Urban Administration Committee (*China Daily* 21 April 2004; Nickum and Lee 2006).

Aqua ex machina? The South-to-North water transfers

After 2010, Beijing plans to be a major beneficiary of a huge supply augmentation project, as the Middle Route 'South-to-North' diversion of Yangtze River is scheduled to be completed. This interbasin transfer, the only one of three that would directly affect Beijing, has been on the drawing board for many decades (for an early set of papers on the project and its potential environmental effects see Biswas et al. 1983) but has only recently received the final go-ahead.

Other elements of Beijing's ecosystem

The focus here has been on water and on documenting Beijing's adaptive responses to shortages both chronic and cyclical. While water is one of the most important elements of the city's ecosystem, it must also be recognized that a number of other 'crises' have become nearly constant elements of life in Beijing, at least for some. Particularly salient are air pollution, energy shortages and land degradation. Air pollution is probably a greater stressor than water on human health, especially within the city proper, while energy shortages may take a greater toll on the economy. All of these elements would seem to follow their own adaptive cycles and are interconnected with each other and with water. For example, months of illegal sand-dredging have transformed the once flat ground of Beijing's largest subsurface reservoir into ravines, directly threatening the water quality of the wellheads and one fifth of the city's water supply. The water source, consisting of thirty-six wellheads around the convergence of Chaobai River in the northeast, supplies water for the Eighth Waterworks of Beijing. The waterworks provides 370,000 m^3 of water for domestic consumption, one fifth of the city's total domestic water supply, to about two million residents every day. It covers 175 km^2 in the southeast part of Beijing. The once flat ground around the reservoir has become a ravine, forty m deep, ten km long and two km wide, with wellheads and high voltage poles towering in between. Some wellheads are almost tottering without sand around the walls, and some high voltage poles have been felled (Xinhua News Agency 2004).

Limitations of time, space and knowledge preclude a more comprehensive treatment of Beijing's ecosystem here, however.

Implications of Beijing's experience

The most important conclusion that comes out of a review of Beijing's water economy over the past few decades is that the city's evolving water institutions are more resilient than one might think from the continuous declarations of crisis and looming disaster of more than the recent past. In particular, Beijing adopted demand management techniques at an early date, even before the decision-making methodologies and analytical frameworks were developed in the late 1980s and 1990s to legitimate them. With limited new sources of water, the relative focus has shifted away from project construction, at least in the near to medium-term. The most effective instruments appear to have been those that relied on command-and-control approaches, supplemented by economic measures. The largest industrial users were identified, audited, monitored and cajoled into reducing their use. The flow of surface water to agriculture was reduced to a trickle with the drastic reallocation of water rights from the Miyun Reservoir, forcing water saving techniques and reductions in irrigation but with little apparent harm to a rural population in rapid economic transition. To date, at least, it would seem that even in an incremental, improvised way, Beijing has been able to achieve a nervous sustainability in using its water supply.

It should be noted that Beijing may present a best case scenario for Chinese cities. As the administrative capital of the country and the residence of China's principal leaders, it has both political clout and administrative capacity. The showcasing of the city to the world in the 2008 Olympic Games has provided it with even more clout, at least in the short-term until Yangtze water arrives.

This does not mean that Beijing always gets its way, however, and it is constantly negotiating with its neighbors over water supply and disposal issues. As all of north China has faced a widening water gap during the prolonged drought of recent years, and as the groundwater has receded to ever deeper reaches, the level of conflict between administrative regions has become increasingly salient. Other cities, such as downstream Tianjin or smaller conurbations with relatively little economic or political influence, may not be so fortunate. Nonetheless, it is worth noting that despite its unfavorable location and much more precarious water supply, Tianjin's economy has also continued to prosper, as has that of the arid Shandong peninsula. The diversion from the Yangtze, now proceeding at full speed after decades of deliberation and the completion of the Three Gorges Dam, is likely to provide a brief respite to the entire region, but will mostly serve to help reduce the demand gap rather than create a new era of expansion.

Still, the maturing water economy perspective appears to help us sort through the complexities of how Beijing's water sector has evolved. We have seen that there is some reason for optimism regarding sustainability, even without huge interbasin diversions.

Acknowledgements

The revival of my interest in Beijing's water economy in the twenty-first century was due in large part to Peter Marcotullio, to whom I am most grateful. At his encouragement, I wrote 'Water and sustainability in Asian megalopolises: The case of Beijing' (2003), which forms the basis for the first part of this paper.

Notes

1. In 1945 public water supply capacity was 9.5 million m^3 per annum. In 1994 total capacity was 1400 million m^3 per annum; of this, about one half was public and the remainder from private and industrial wells (Sternfeld 1997: 132, 138).
2. Grain rebounded to 700,000 tons in 2004 and 949,000 tons in 2005, due to a major push to increase the area of cropland from 154,000 ha to 192,000

ha, and possibly at the cost of a drop in higher value farm output, such as vegetables and meat (*People's Daily Online*, at *http://english.people.com.cn/200601/31/print20060131_239467.html*, viewed 10 August 2006).

References

Beijing Municipal Water Resources Bureau (1999), *Beijing's Flood and Drought Disasters* (in Chinese), Beijing: Water and Power Press.

Beijing Shi Renmin Zhengfu and Zhonghua Renmin Gongheguo Shuilibu (2001), *21 shiji chuchi (2001–2005 nian) shoudu shui ziyuan kechixu liyong guihua zong baogao*, Beijing: Shuili Shuidian Chubanshe.

Beijing Water Resources Bulletin 2001 (in Chinese) [n.d.], Beijing: Beijing Water Resources Bureau.

Biswas, A., D. Zuo, J. E. Nickum and C. Liu (eds) (1983), *Long Distance Water Transfer in China*, Dublin: Tycooly International.

China Daily (2003), 'Water diversion helps ease Beijing thirst', October 3, *http://www2.chinadaily.com.cn/en/doc/2003-10/03/content_269165.htm*

Hu, C. and S. Nie (1992), 'Dachengbao xiezouqu' (A contracting concerto), *Zhongguo shuili*, 8: 3–8.

Li, J. (2004), 'Beijing raises water prices, ninth in 13 years', *China Daily*, August 2, *http://www.chinadaily.com.cn/english/doc/2004-08/02/content_356772.htm*

Lu, Z., J. Sun and L. Lu (eds) (2002), *Beijing hanqing yu duice* (*Beijing's drought conditions and policy responses*), Beijing: Meterology Press.

Ministry of Agriculture (2004), *China Agriculture Statistical Report (2003)* (in Chinese), Beijing: China Agriculture Press.

Nickum, J. E. (1997), Beijing's maturing socialist water economy, in J. E. Nickum and K. W. Easter, *Metropolitan Water Use Conflicts in Asia and the Pacific*, Boulder: Westview: 37–60.

Nickum, J. E. (2003), 'Water and sustainability in Asian megalopolises: The case of Beijing', in Saskia Sassen (ed.), *Human Settlement Development*, Encyclopedia of Life Support Systems (EOLSS), developed under the auspices of the UNESCO, Oxford: Eolss Publishers,

Nickum, J. E. and K. W. Easter (1997), *Metropolitan Water Use Conflicts in Asia and the Pacific*, Boulder: Westview.

Nickum, J. E. and Y. F. Lee (2006), 'Same longitude, different latitudes: Institutional change in urban water in China, north and south', *Environmental Politics*, 15, April: 231–247.

Peisert, C. and E. Sternfeld (2005), 'Quenching Beijing's thirst: The need for integrated management for the endangered Miyun Reservoir', *China Environment Series*, 7: 33–45.

Sternfeld, E. (1997), *Beijing: Stadtentwicklung und Wasserwirtschaft*, Berlin: Technische Universität Berlin.

Wang, H., D. Qin, J. Wang, L. Luo and Y. Pei (2003), *The Rational Disposition of Water Resources in the Huang-Huai-Hai Region*, Beijing: Science Press.

Xinhua Net (2005), *Beijing lianxu 3 nian shijian meinian jieshui 1 yi lifangmi*, October 28, *http://news.xinhuanet.com/politics/2005-10/27/content_3692223.htm*

Xinhua News Agency (2004), 'Illegal sand-digging threatens one fifth of Beijing's water supply', November 6, viewed November 11, 2008 at *http://english.peopledaily.com.cn/200411/06/eng20041106_162980.html*

Yearbook of China Water Resources 2003 (2003) (in Chinese), Beijing: China Water and Power Press.

12
The Linkages Between Poverty, Environment and Transboundary Water Management in Southwest China's Yunnan Province

Katri Mehtonen

Introduction

China has the greatest water resources in the world with renewable resources of some 2812.4 billion m^3. However, the per capita figure is only about 2300 m^3. Additionally, water is unevenly distributed both geographically and seasonally. By the year 2030 the per capita possession of water is expected to drop to 1760 m^3, which means China will be a country short of water (He et al. 2003). Among the provinces and autonomous regions in China, the top five in terms of per capita water resources are in the western region, namely Tibet (206,892 m^3), Qinghai (13,978 m^3), Yunnan (5953 m^3), Xinjiang (5774 m^3) and Guangxi (4415 m^3). Consequently, these regions are considered as the strategic resource base for China's water security and sustainable development (He et al. 2003).

Yunnan Province is situated in Southwest China. It is one of China's border provinces, neighboring Myanmar, Laos and Vietnam. The province is very mountainous: the share of mountains of the total area of 383,400 km^2 is about ninety-four percent, while the remaining six percent is covered by small-sized flat basins. The altitude in the province varies between seventy-six and 6740 m within a short distance. Many great transboundary rivers flow through Yunnan and thereafter enter the neighboring countries. These rivers include the Mekong, Salween, Irrawaddy, and the Red River. Of these, the latter two also originate in Yunnan. Even though total water resources in Yunnan are abundant with a figure of about 250 billion m^3, water availability is not obvious due to the mountainous topography and deep river gorges. Most of the fields are located high in the mountain slopes, while the rivers flow down in the gorge, making it hard to access the water (He and

Figure 12.1 Yunnan province and the international rivers flowing through it

Source: He and Gou (2003).

Liu 2002). Therefore, many farmers actually suffer from lack of adequate irrigation water. Additionally, the timely distribution of water is uneven as the greatest need appears during the dry season.

The transboundary rivers flowing southwards from Yunnan make the region particularly important in terms of cross-border impacts. However, China's participation in international cooperation over transboundary water resources has been considered significantly inadequate by several of the downstream countries. Development on the Mekong River within China (Lancang in Chinese) has in particular raised many concerns. The Mekong flows through six countries, including China, Laos, Myanmar, Thailand, Cambodia and Vietnam. China is not a member of the Mekong River Commission (MRC), and has been accused of causing damage downstream by the large dams that are being built on its side of the river. There are other issues as well, such as the navigation improvement project of the upper Mekong, based on an agreement among the four upstream countries (China, Laos, Myanmar and Thailand), and the plans for the upper Salween River.

According to Chinese law, the planning and management of large hydropower dams falls directly under the central level of governance. Thus, the upper Mekong dam cascade is supposed to be a national-level project. However, talking only about China as a whole in terms of the transboundary river scene is misleading. In recent years the provinces have gained more power, following the decentralization process. Yunnan province obviously plays some kind of a role in the decision-making process over the river projects in its territory. What this role exactly is has not been clearly articulated, at least among international scholars. It seems that the central government will not implement projects that are clearly against the will of the provincial administration, but this impression has not been confirmed. Based on the complicated decision-making structure, it is difficult to say how much influence local issues and conditions within Yunnan have on the river development.

Yunnan is a poor province both in national and international terms. Moreover, the demanding physical environment adds to the challenges of poverty alleviation. One can hypothesize, then, that tackling these issues requires a major part of the provincial government's capacity and resources. Poverty issues in the national border regions appear to be the most difficult to tackle, mainly due to the remote location and poor infrastructure. However, improving the infrastructure risks causing environmental degradation, which could in turn have its own negative effects on local livelihoods. Building hydropower dams and international transportation routes has in particular been seen as a major threat to the regional ecosystem (AIRC 2004). The needs for such infrastructure are related not only to local development but also to China's economic regionalization with Southeast Asia. Particularly the Greater Mekong Subregion (GMS) program and the China-ASEAN free trade zone development are driving forces for infrastructural improvements.

It is difficult to prevent environmental damage in a region that is desperately seeking a way out of poverty. The lack of financial, technical and scientific resources hampers conservation efforts. Moreover, in the case of transboundary resources such as international rivers, the effects across the borders should also be considered. The key issue raised in this chapter is whether the poverty and development challenges in Yunnan require such an effort that it would not be realistic to increase consideration of transboundary impacts. This study is framed mainly to discuss issues within Yunnan province, although some aspects of the central government's contribution are touched upon as well.

Objectives

This chapter discusses the conditions in which transboundary river projects are undertaken in Yunnan province. It also examines how China's development policies have responded to the great challenges facing the region. Of particular interest is how these adverse conditions reflect transboundary issues. More precisely, how do the challenges affect transboundary impacts caused by poverty-driven development in Yunnan? In the following, some aspects of the question are discussed, paying special attention to transboundary river management. Additionally, an overview is carried out of the environmental and poverty alleviation projects being implemented in Yunnan. The aim is to highlight the significant efforts made by the multiple stakeholders to mitigate these conditions.

Section two introduces the transboundary river issues present in the study region. Section three discusses the poverty and environmental challenges within Yunnan province. In section four, some policies for poverty alleviation, economic development and environmental protection in Yunnan are presented. Section five analyses the potential reflections of the challenges in Yunnan on the transboundary river issues. Some suggestions by the author are presented in the final remarks.

Transboundary river issues

On a global scale, China ranks first in terms of its number of international rivers and quantity of transboundary water resources. Of the fifteen major international rivers in China, eleven are situated in the western region. The southwest provinces of Tibet and Yunnan in particular have the highest concentration of international rivers. This region contains the headwaters of the major Asian international rivers, such as the Ganges, Brahmaputra, Yangtze, Mekong and Salween, and is thus regarded as the 'water tower' of Asia (He et al. 2003).

Due to abundant flow and steep topography, the exploitable hydropower potential in Yunnan is huge, accounting for some 90,000 MW (Piyasvasti 1999). Noting that China is facing increasingly serious water and power shortages, it is not difficult to imagine that hydropower development in Yunnan is seen as one of the major solutions. With an annual economic growth of eight to ten percent during the past five years, the country is short of energy supplies, in particular electricity (BBC News Online 2004). Therefore, the central government plans to triple the installed hydroelectric capacity from current levels to 270,000 MW by 2020 (IRN 2004a). Moreover, additional hydropower is expected to allow the closure of some highly polluting coal-fired plants, as well as avoiding constructing new ones (ADB 1998a; 1998b).

China's plans for the transboundary rivers include the ongoing construction of seven or eight large dams on the upper Mekong and the planned dam cascade on the Salween River. Additionally, Vietnam has suggested that China dam the upstream of the Red River to prevent sedimentation problems downstream. The Mekong project in particular has raised strong international opposition while the Red River case is the only one in which a downstream country openly wishes a dam to be built upstream. In the case of the Mekong, the four downstream countries have expressed serious concerns about the dams' negative impacts. However, Thailand has shown willingness to support some of the schemes through the Electricity Generating Authority of Thailand (EGAT). As one of the three riparian countries of the Salween, Thailand is also planning joint hydropower development with Burma on the Burmese side of the Salween. Additionally, Laos is developing hydropower on the Mekong tributaries. The usual image of China as the only serious dam developer in the region is therefore far oversimplified.

Both the promoters and the opponents of China's dams seem to lack scientific bases for their arguments, as nobody has been able to assess the environmental and social impacts within the whole river basins. The obstacles in conducting such an assessment include the scale of the research required, problems in information sharing and the reluctance of certain Chinese stakeholders to perform such studies. Additionally, due to China's ongoing transformation from a planned economy to a market system, filling the research gaps on international rivers began only relatively recently during the reform and opening-up period that started in 1978 (He et al. 2004).

In the following, the ongoing and planned dam constructions on the transboundary rivers within Yunnan are briefly presented.

Mekong

The Mekong is the ninth largest river in the world if measured by runoff and the largest water resource in Southeast Asia, with an average discharge of 15,000 m^3/s. From its total length of 4880 km, 1161 km of the river is situated in China. The Mekong River basin covers densely populated areas with sixty million people depending directly on natural ecosystems for their livelihoods. Fish catches are the major sources of protein throughout the basin and the river is important for rice production particularly in the Mekong Delta in Vietnam.

Damming of the Mekong started in 1995 when the Manwan dam with an installed capacity of 1500 MW was completed. The Dachaoshan dam began operation in 2003. Xiaowan, the largest dam in the cascade with a dam height of almost 300 m and capacity of 4200 MW, is under construction and is planned to start operating in 2013. The Xiaowan system is supposed

Figure 12.2 Upper Mekong dam cascade

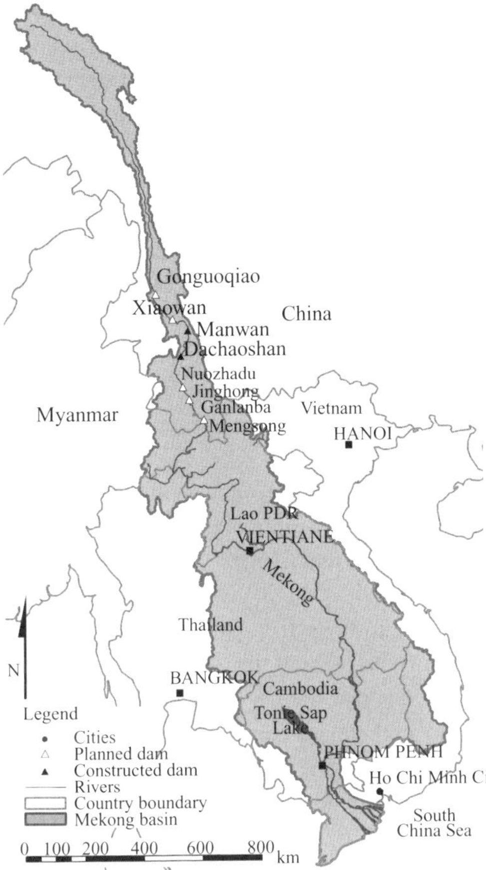

Source: Kummu (2004).

not only to provide energy, but also to act as a regulating reservoir for downstream dams. Additionally, it is expected that it will help with the serious sedimentation problems at the existing dams by trapping most of the sediments. The 1500 MW Jinghong dam is also under construction. Another huge dam, the Nuozhadu, will provide 5000 MW and is expected to be in operation in 2017. The current plan involves altogether seven or eight large dams, with a total installed capacity of about 15,000 MW (see Figure 12.2). The majority of the electricity provided by the cascade will be transferred to the regions east of Yunnan, particularly to Guangdong, and only a minor part will be consumed locally. Additionally, some of the electricity will most probably be exported across national borders, mainly to Thailand.

The Mekong has six riparian countries and particularly the four downstream countries (Laos, Thailand, Cambodia and Vietnam) have expressed serious concerns about the environmental and social impacts of the dams. The major worries center on the impacts of altering the natural regime of the river (Dore and Yu 2004), especially since, as noted earlier, no environmental or social impact assessments covering the whole river basin have been made. In fact, to date China has not performed any real negotiations with the other riparian nations over the project. China's main argument (referring here to the official view) is that if there will ever be any downstream impacts, they will be mainly positive. The dams will apparently make the flow variations on the river smaller by increasing the dry season flow and decreasing that of the wet season. Therefore, the proponents emphasize that the cascade will help alleviate the flooding and drought problems downstream. On the Mekong delta in Vietnam, for example, the increased dry season flow would mitigate the saline water intrusion problem.

However, the dam proponents do not mention that increasing dry-season flows and decreasing wet-season ones will probably hamper downstream ecosystems and some important livelihood forms. The river gardens along the riverbanks will be flooded if the dry season flow increases. The dams will block the passage of migratory fish and thereby decrease the amount of and change the variety of fish. Overall, the changes in the flow regime will cause serious problems to the fish, which have adapted to the natural variations. Additionally, blocking the upstream sediment flow is a serious issue, which has often been neglected. One of the affected regions will be Cambodia's Tonle Sap Lake, which is a unique lacustrine-wetland ecosystem and the Mekong's major natural reservoir with annual water level fluctuations of eight to ten meters. The Tonle Sap River, which usually flows to the lake from the Mekong, reverses its direction during the flood peak. The sediment flux to the lake in the flood season is many times larger compared to the outflow flux in the dry season. It is hypothesized that sediment carried by the Mekong waters to the Tonle Sap Lake bring in essential nutrients that feed into the lake's food webs (Sarkkula et al. 2004; Sarkkula et al. 2003; WUP-FIN 2003). Upstream developments like the construction of dams have already led to significant trapping of sediment (Kummu et al. 2004) and nutrients and may reduce the fertility of the Tonle Sap system. Considering all the aspects, it is difficult to accept the argument that flow regulation will only have beneficial impacts.

Part of the problem is that China is not a member of the Mekong River Commission (MRC), which is a cooperative institution between the four downstream countries. The MRC was established in 1995 for the sustainable development, utilization, conservation and management of the Mekong River

Basin water and related resources. Basically, all major plans on the Mekong mainstream should arguably go through a process within the commission and its member countries. Even though China's involvement with the MRC has slightly increased during recent years, there seems to be reluctance for full cooperation coming from both sides.

In terms of the people living on the poverty margin in Yunnan, the essential question is to consider to what extent the dams will benefit the local people. In view of the topic of this chapter, the hydropower projects could be more easily justified if it could be shown that they were an integral part of poverty alleviation in Yunnan. Yet, as the dam developers have thus far been reluctant to make the figures of electricity and profit allocations public, the suspicion remains that the local people will not significantly benefit. The operation manner of the cascade in relation to the natural flow patterns is another issue which has not been adequately discussed. Now that the building of the dams has already begun and seems to be moving forward, the operation rules should be prepared through a participatory process, including the downstream countries. It would be useful for all the stakeholders to focus on this issue, rather than to continue arguing whether the dams are more beneficial or harmful.

Salween

Unlike the Mekong and despite its huge hydropower potential, the Salween River remains relatively untapped. It is actually one of the last pristine rivers in Asia and in the world. Plans to develop the river are nevertheless in the discussion phase between Thailand and Myanmar, and in China. The Salween originates from the Qinghai-Tibet Plateau, and runs through the Yunnan province into Burma. About 1370 of the total stretch of 2400 km is situated in China, whereas the river forms the Myanmar-Thailand border for only about 110 km. The total catchment area is about 320,000 km^2, of which fifty-three percent is situated in China, forty-two percent in Myanmar and five percent in Thailand.

China's plan includes building thirteen dams on the Salween, mainly within the Nujiang Prefecture of Yunnan province (Figure 12.3). The total capacity of the cascade would exceed 21,000 MW (*Yunnan Daily* 2003a). Particularly the local governments in Yunnan support the dams. They claim that hydropower will help the poor region and ethnic minorities to alleviate poverty and promote social and economic development (*Yunnan Daily* 2003b). Furthermore, hydropower is seen as a clean power source and increasing its share would help to move away from heavily polluting coal-based power (*BBC News Online* 2004).

Figure 12.3 The planned Upper Salween cascade

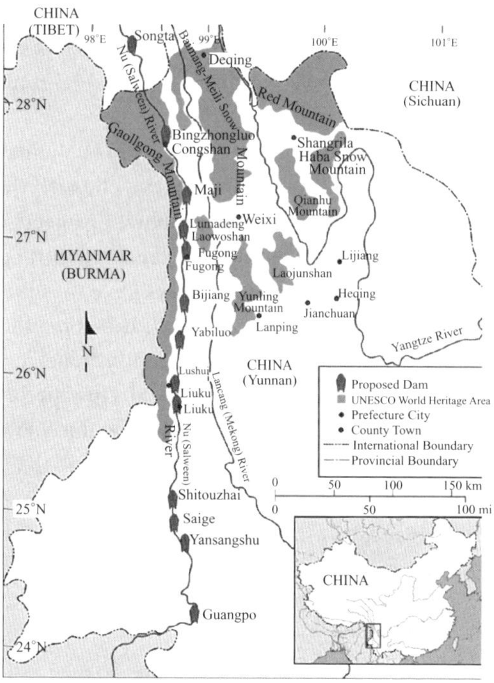

Source: IRN (2004a).

The Salween plan raised broad environmental awareness and public opposition unprecedented in China, and even provoked rifts within the government (*New York Times* 2004b). The main arguments of the dam's opponents include environmental destruction, human suffering and the lack of a proper impact assessment. For example, the area in China called the Three Parallel Rivers where the Yangtze, Mekong and Salween run side by side was designated as a UNESCO World Heritage Site in 2003. Additionally, it was claimed that there has not been any consultation with neighboring countries and affected communities (*New York Times* 2004b). The debate over the project has been strong among environmentalists, government officials, power companies and concerned academics. Additionally, UNESCO expressed concerns over the impacts to the World Heritage site and requested a proper environmental impact assessment (IRN et al. 2004). The project, which was originally scheduled to start in September 2004, was halted due to the strong opposition. In April 2004, Chinese Premier Wen Jiabao ordered that a major review of the project should be conducted before implementing the plan (*New York Times* 2004a). Dam proponents are reputed to be pushing for the construction of a slimmed-down version of the original project with four dams instead of the originally proposed thirteen (SCMP 2004). For the time being, at least, the project remains in abeyance (McDonald 2008)

Red River

The Red River originates in Yunnan Province and then flows through Vietnam to enter the South China Sea. The total length of the river is 1280 km, out of which 667 km is in China. More than half of the total catchment area of 113,000 km^2 is situated within China. The discharge of the Red River is highly variable, from a minimum recorded dry season discharge of 370 m^3/s, to a maximum of 38,000 m^3/s. The continuous floods in the river basin, caused by great amounts of river-borne silt, changed land-use practices and deforestation, pose a significant problem especially in the Vietnamese Delta Region. Therefore, upstream development might benefit Vietnam in terms of reduced siltation.

China and Vietnam share some common interests in developing the Red River. China could produce hydropower and Vietnam hopes for more reliable flood protection and soil erosion control. Decreased sedimentation would help the downstream areas where serious floods occur regularly and threaten both human lives and property. The topic was discussed in 1998 when an international symposium was held between the two countries (He

Figure 12.4 Red River Basin

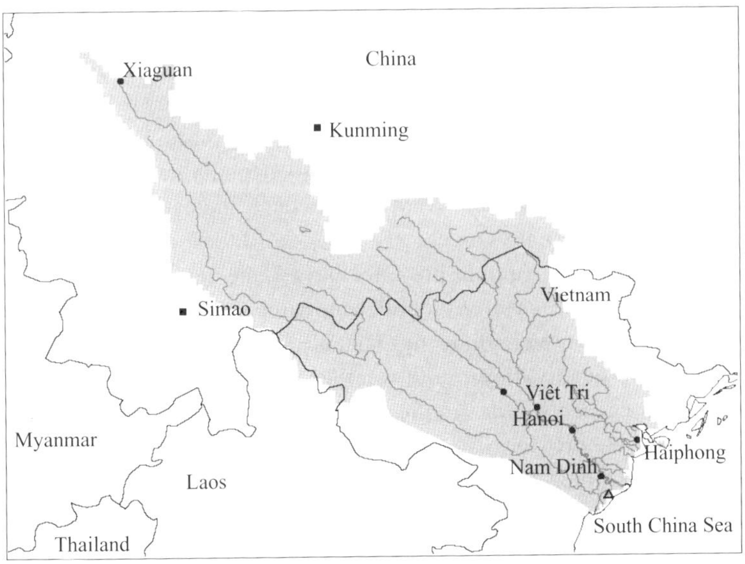

2004). In the symposium, Vietnam suggested that China build a dam on the Chinese side of the river to help navigation, decrease flooding and increase irrigation water. So far the only dams on the river are small ones on the Red River branches.

Sino-Vietnamese relations seem to have improved remarkably in recent years, following a turbulent past of conflicts and political disputes. In 2001, Yunnan agreed to offer Red River water levels and flooding information to Vietnam (He 2004). The two countries also share similar interests in the trade sector and it seems likely that bilateral cooperation will increase in the future, including some projects on the Red River.

Poverty and environmental challenges in Yunnan

Poverty is indeed a serious issue in the region. Within Yunnan, there are forty-two counties classified as national-level poverty counties and seventy classified as provincial-level poverty counties (the commonly used poverty line in China is 635 *yuan* per capita per year). Additionally, there are a number of very poor townships. The major obstacles for poverty alleviation include remoteness, challenging physical environments and poor infrastructure. Some of the poorest border regions in Yunnan have additionally suffered economic damage both during the Vietnam War and the Sino-Vietnamese war in 1979–1983. China's membership in the World Trade Organization (WTO) is considered as a new threat for the still very old-fashioned agricultural production in Yunnan. The WTO accession of 2002 is expected to decrease the already low prices of agricultural products and thus to hurt the local economies at least in the short-term. The Great Western Development Strategy, launched in 1999, is supposed to bring more prosperity to the western regions. However, the strategy also strongly emphasizes natural resource and energy transfers from west to east, indicating possibly more benefits to the eastern provinces.

The watersheds in Yunnan province have some special characteristics, which influence the environmental conditions and livelihoods of local people. These features include significant elevation variations with high mountains and deep valleys, cold weather in the highlands and plentiful precipitation caused by the southeast and southwest monsoons (He et al. 2001). The whole province is at an average elevation of about 2000 m. The vast land is inhabited by a relatively sparse population, a significant share of which still lives in extreme poverty.

Yunnan has in many cases acted as the representative of China in discussions over the use of China's transboundary rivers. Nevertheless, it remains unclear as to how much the province really has to say on the level of cooperation and the plans for these rivers, and how much is decided at the central state level. The hypothesis of this study is that in any case the province's poor economic performance has at least something to do with the dam projects. It is therefore necessary to study the actual situation in Yunnan. Two factors were selected for closer examination due to their apparent significance in the region: poverty and environmental issues. These factors are particularly important, because the river basins in the region are characterized by serious tensions between economic development and environmental conservation (AIRC 2004).

Poverty

By the end of 2003, China's poverty-stricken population was thought to be twenty-nine million, out of which most people live in the western parts of the country (*China Daily* 2004). The figure has come down from 250 million poor people before the reform and opening drive began in 1978, but in the 1990s the pace of poverty reduction clearly slowed down (ADB 2004).

Despite the rather rapid economic growth during recent years, Yunnan remains one of the poorest provinces in China. GDP in the province grew by 5.4% during 1999–2000, reaching US$ 23.62 billion in 2000 (*China Economic Review* 2002). However, during the same year GDP per capita remained at $US560.1, which is very low both by international and Chinese standards (transformed from *yuan* to $US according to 2000 rates). Half of Yunnan's population earns less than $US80 a year (*New York Times* 2004b) and unemployment is a remarkable problem. By the end of 2000, the poor population (earning less than 635 *yuan* per capita per year) in Yunnan was about 10.05 million, and the population suffering from inadequate food was about 1.5 million (He et al. 2002). The majority of the poor live in the upstream area of the great rivers and not only suffer from frequent natural disasters and fragile ecological environments, but are also seriously affected by the national logging ban initiated in 1998. Additionally, access to water resources is difficult due to the scattered flatland cut by deep gorges and consequently, the costs of irrigation facilities are very high.

Like the other western regions of China, Yunnan has lagged behind the development of the east coast and the capital region. The strongest contributor may be the earlier state policy designed to mainly support certain regions, which were then expected to pull the rest of the country along with their development. However, Yunnan's remoteness and natural environment also have a part to play. The province is landlocked and situated far from the coast and major economic centers. The mountainous topography is challenging for any production form. Adding the high-altitude monsoon climate to remarkable regional differences and elevation changes, it is not surprising that transportation is difficult and maintaining roads and railways requires significant efforts. Every year floods cause destruction, and particularly in peak-flood years there is serious damage and loss of roads, bridges, houses, fields and even human lives. Moreover, the floods have some serious side effects, among others landslides and mudflows, which cause damage on their own and further worsen the flooding. Both the central and provincial government support disaster prevention and reconstruction of the lost infrastructure, but the money is not sufficient to stop the flood-affected people's income from falling dramatically.

Yunnan is also home to the richest variety of ethnic minorities within China, accounting for twenty-six out of fifty-two different groups at the national level. The total population of Yunnan is about 42.9 million people (*China Economic Review* 2002), of which the share of minorities is more than 13.6 million. Several minorities live in the border regions and some groups actually live on both sides of the national borders with Myanmar and Vietnam.

Many of the areas where the majority of rural poor in China live share common features (He et al. 2002). Yunnan falls into this category, both in terms of a large poor population and features shared with other impoverished rural areas, including:

- location in the middle or western part of the country
- mountainous areas with high elevations
- unfavorable physical (topologic, geomorphologic and geological) conditions
- limited and infertile farmland with poor water access
- uneven distribution of precipitation and water resources
- fragile and deteriorating eco-environment
- limited infrastructure and widespread power shortages.

The linkages between environmental challenges and poverty

Poverty in Yunnan is strongly related to the environmental conditions and the very challenging physical environment, resulting in the frequent occurrence of natural disasters. The cause of the disasters lies partly in the heavy rainfall during the wet season and the reduced forest cover, leading to soil becoming unstable. Approximately 80,000 people have been killed and forty-five billion *yuan* of direct economic losses incurred in Yunnan during the past five decades because of disasters such as landslides and debris flows (He and Liu 2002). Moreover, over forty percent of the land is endangered by soil erosion, and the annual sediment run-out of the province is about 3×10^8 t (He et al. 2003). The Red River basin in particular faces serious erosion with a long-term average sediment concentration of 0.36–4.93 kg/m^3. As a result of deforestation and soil erosion, the loss of soil fertility causes serious problems in Yunnan. About 518 million tons of soil is lost each year, and 2442 reservoirs and hydropower stations are suffering from consequent sedimentation (Yang cited in He and Liu 2002).

In terms of biodiversity, Yunnan supports about half of the plant and animal species found in China. Even though the state of the environment in Yunnan is relatively good, the loss of forest cover, infrastructure building, pollution and climate change are causing some negative impacts. As one example of a disturbed ecosystem, the Dianchi Lake on the southern side of the provincial capital Kunming is seriously polluted. More than $US two billion has been spent to clean up China's sixth largest lake since 1993. Sewage treatment plants were built and in 2000, about sixty percent of the city's sewage water was treated (US Embassy, Beijing, 2000). Still, there are large quantities of untreated wastewater flowing into the lake from the surrounding rivers. Even more seriously, agricultural activities keep polluting the water due to the heavy use of chemical fertilizers and pesticides. For these reasons, pollution-control efforts have had only limited effect and the pollution levels in some parts of the Dianchi still exceeded the fifth level (the worst) on China's water quality scale in 2000.

Pressures to alleviate poverty and to develop economically often contradict environmental conservation objectives. Alternatively, the natural environment has a particularly strong relation with human welfare in Yunnan. In particular, the many ethnic minorities living in the region are strongly dependent on the environment and its capacity to support their livelihoods. In addition to the cultivated products, wild plants are an important part of the diet in many regions. Moreover, in many regions locally procured herbal medicine is the only medical care rural people can afford. If there were any serious environmental damage, the poorest people would clearly lose their only source of food, income and health support.

Environmental degradation is a difficult matter to avoid in the poverty-stricken province. For example, strictly defining permitted industrial pollution levels usually hampers production, and local development strategies might not allow this to occur. It seems that at all levels of governance, starting from the centrally administered State Environmental Protection

Administration (SEPA), the environmental departments lack adequate power to effectively influence other departments' operations. One reason is that environmental governance has not yet fully stabilized itself in the Chinese system. Moreover, local departments are sometimes in a difficult position, as they have to respond both to the corresponding central level department and to the local government (Makkonen, 2003).

Policies for poverty alleviation, economic development and environmental protection

Multiple stakeholders are endeavoring to alleviate poverty in Yunnan and particularly to help rural people to develop their economies. Both the central and provincial governments have allocated support and preferential policies to the specifically defined poverty counties in Yunnan. China is also paying growing attention to environmental issues and sustainable development. Several academic institutions are currently studying issues related to the environment and ecological security. Moreover, there are several non-governmental projects being carried out in Yunnan, aiming at grassroots-level improvements in the lives of local people. In the following, some of the planned and ongoing projects related to poverty alleviation and environmental conservation within Yunnan and China's western regions are highlighted.

Key counties for poverty reduction

The central government has some nationwide policies for poverty alleviation, including the definition of national level poverty counties. In Yunnan, there are forty-two such counties, called the key counties for national poverty reduction and development work. This geographically targeted approach to rural poverty reduction has continued since 1986 (ADB 2004). The poor counties have been eligible for subsidized loans, food-for-work programs and special development funds. The program covers about 592 of China's counties, out of a total of 2074. Additionally, the provincial government has defined seventy counties in Yunnan as provincial level poverty counties (including the forty-two national poverty counties).

Nevertheless, between 1997 and 2000 the poverty reduction rate in national poverty counties was lower than in the other Chinese counties (ADB 2004). This can be partly explained by the fact that the income distribution patterns favor eastern over western provinces and cities over rural populations. During the same period, the GDP growth in the eastern provinces produced considerably more rural poverty reduction than in the western provinces.

Developing the west

After first promoting development in the eastern regions for a long time, the central government has changed its focus into developing the west of the country as well. The western development strategy was launched in 1999 with the aim of supporting the relatively backward and poor western areas to catch up with the east. Even though this region occupies two thirds of China's territory and one third of its population, its share of the national economy is currently less than one fifth and the GDP less than half of that in the east (*China Daily* 2004a).

The western region is defined to cover eleven provinces, autonomous regions and municipalities, including Shaanxi, Qinghai, Sichuan, Yunnan, Guizhou, Ningxia, Xinjiang, Inner Mongolia, Gansu, Tibet and Chongqing (Figure 12.5). The reasons for lagging economic development include water shortages, poor education and public health facilities, and the shortage of educated personnel.

The western development strategy consists of a series of policies preferential to economic development, including investments, preferential tax rates and flexible policies. Energy, communications and water conservancy projects are prioritized in particular (*China Daily* 2004b). The leading ideology is to better coordinate development between the economy, society and environment, and thereby to promote scientific development in the region.

Environmental degradation is severe in many parts of the western provinces. In this region where most of China's major rivers originate, there are serious problems with soil and ground water pollution, soil erosion and water shortages. The ecological initiatives include, among others, reconverting farmland to forests and pasture, planting trees, and preventing and controlling industrial pollution. Additionally, efforts are being made to restrain industrial development and reduce the use of natural resources. However, this aim seems at least partly to conflict with the economic development aims.

Since the launch of the strategy, a total of US$ 100 billion has been allocated for water conservation, communications, environmental protection, compulsory education and healthcare. In addition to public investments,

Figure 12.5 China's western region, shown in the grey color shades

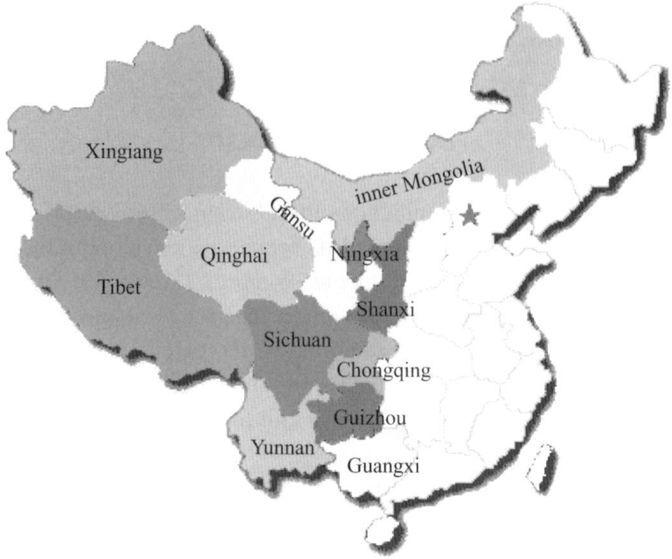

Source: Office of the Leading Group for Western Region Development of the State Council (2004).

non-governmental investors are encouraged. Additional loans to projects in infrastructure and natural resources exploitation will be granted for the region in the future (*China Daily* 2004c).

In addition to the operations within the western regions, the key projects concern natural resources sharing and trade increase between the west and east. These projects include the west-to-east electricity transmission and natural gas pipeline, highway extensions and the Qinghai-Tibet Railway. During the first half of 2004, about 600 MW of power generating capacity was installed in the west and the transmission lines were extended by 2400 km (*China Daily* 2004a). The gas pipeline from the Tarim area of the Xinjiang Autonomous Region to Shanghai was completed ahead of the planned schedule. When one looks at the key projects it is easy to note that the infrastructure development will greatly benefit the eastern parts of the country.

Southwest China, including Yunnan, Sichuan, Guizhou and Guanxi provinces, Chongxing municipality and the Tibet autonomous region, is particularly rich in water resources with a potential of generating 256,740 MW (37.9% of the national total). This region therefore plays an important role in the west-to-east power transfer plans (He and Liu 2001). The Xiaowan dam on the Lancang River, for example, is expected to be one of the key outcomes in implementing the power transfer program (Ma Hong 2006).

Ongoing projects in Yunnan

The following section introduces some of the ongoing and planned projects for poverty reduction and environmental protection in Yunnan. The purpose is not to cover the whole variety of such projects, but to highlight some of the valuable efforts made to improve the situation within the province.

Ecosystem changes

In 2004 a group of Chinese academic institutions launched an eco-security project within Yunnan, called 'Ecosystem Changes in the Longitudinal Range-Gorge Region (LRGR) and Transboundary Ecological Security in Southwest China 2004–2008' (AIRC 2004). The project deals with changes in the natural environment and biodiversity, and additionally looks at the transboundary ecological issues on both sides of the national borders. The longitudinal range-gorge region encompasses the major transboundary river basins within Yunnan, including the Red River, Mekong, Salween and Irrawaddy. The region is extremely rich in biodiversity and provides an important ecological and economic corridor linking China and Southeast Asia. However, due to hastened economic development and natural resources utilization, there have been rapid and drastic changes in the ecosystem and species diversity. Additionally, different development programs, such as China's Western development strategy, the Greater Mekong Subregional (GMS) program and the proposed China-ASEAN free trade zone, threaten to disturb the delicate ecosystems.

The Ecosystem Changes project, named as one of the national key projects for basic research, will focus on the corridor-barrier phenomenon and its effect on ecosystem changes. Barriers in this context mean extreme natural barriers, such as mountains, and corridors signify natural passageways, particularly rivers. The topics of investigation include patterns in ecosystem change, quantifiable changes in ecosystems, driving factors and ecological security thresholds. One target is to create an evaluation framework for ecosystem diversity. Interactions between major engineering projects and the environment will be investigated, transboundary ecological impacts of ecosystem change studied, and mechanisms for sustainable development that maintain transboundary ecological security formulated. The overall goal of the project is to identify ecosystem changes within Southwest China and their transboundary impacts in the GMS. The expected results include mechanisms for maintaining ecosystem diversity, evaluation indices for the ecological impacts of engineering projects, and methodologies for maintaining ecological security across political borders. These results are supposed to facilitate sustainable implementation of international programs, to help in developing a scientific basis for environmental restoration, and to facilitate environmental diplomacy. The final aim is to develop strategies for maintaining ecosystem functions and conserving species diversity.

The project is funded by the Chinese Ministry of Science and Technology. Because of the funding base, the study is framed mainly to consider issues within China and in the fields of natural sciences. However, an additional component covering social issues has been suggested by the Asian International Rivers Center (AIRC) and its partners (project proposal, unpublished). The 'Improved Transboundary Environment Governance Through Participatory Joint-fact Finding and Collaborative Management' component would expand the Ecosystem Changes project by integrating social science research into the natural sciences, bringing policy and management stakeholders into the research and by engaging actors from the other GMS countries. The purpose of the sub-project is to strengthen governmental and civil society institutions and policies for transboundary resources management.

Impacts of trade liberalization

One of the ongoing projects deals with the impacts of trade liberalization on the environment and the rural poor in Yunnan (AIRC 2004). The study is being carried out by the Asian International Rivers Center and WWF (World Wildlife Fund) China, and is funded by the World Bank. It focuses on the communities surrounding the protected areas in the Red River basin, along the China-Vietnam border.

The research will particularly explore the impacts of the WTO membership on corn growers in the region. It is expected that as protections against foreign grain imports decrease in China, grain prices will drop, increasing the pressure on local farmers to move production into protected areas to maximize their crop yields. In terms of environmental impacts, the changes to land use, water quality and biodiversity will be studied. Additionally, the wellbeing of local communities in one of China's poorest regions is under investigation.

Integrated ecosystem and water resources management

The 'Integrated Ecosystem and Water Resources Management of Lancangjian (Upper Mekong) River Basin: A Pilot Study in the Fengqing and Xiaojie Watershed' is a joint Sino-Dutch study between several Chinese institutions (e.g. Yunnan University, Yunnan Bureau of Hydrology and Water Resources, Yunnan Forestry Department) and Dutch counterparts, led by the UNESCO-IHE Institute for Water Education. The time scale of the project was 2003–2005.

Like the Ecosystem Changes project, this study focused on changes in the ecosystem, mostly caused by increasing human pressure. The study region consisted of two ecosystems situated in the upper Mekong system: the Fengqing and Xiaojie catchments. The four focal points of the project included soil, water and biodiversity conservation through ecosystem restoration; assessment of slope stability and landslides hazard; water resources assessment; and integrated watershed planning and management. The project particularly investigated the water ecosystem impacts, such as soil erosion, landslides, increased sediment loads, habitat fragmentation and species loss. The purpose was to find ways to develop an Integrated River Basin Management (IRBM) approach. The aim was to identify sustainable development scenarios that would maintain the integrity, diversity and symbiosis of local livelihoods, cultures and ecosystems. Thereafter, policy-makers and water resources managers were to be given insight into the economic, social and ecological effects of the different scenarios. Additionally, they would be provided with practical guidelines for a more sustainable development of the selected catchments.

Sustainable development

According to current regulations, the Chinese provincial governments need to prepare a sustainable development plan. The 'Yunnan Sustainable Development Action Plan' (YSDAP) is a project aimed to assist the Yunnan provincial government to prepare a sustainable development plan for the province. The assisting institute is the Yunnan Environmental Development Programme (YEDP), a grant program funded by the Department for International Development (DFID) of the British Government. The aim of the YEDP is to promote environmentally sustainable development and poverty alleviation in the Yunnan province. The counterpart from the provincial government is the Yunnan Development and Reform Committee.

YEDP's work to strengthen the YSDAP consists of two phases. The first phase will provide government officials with training on six key tools for helping to develop a sound action plan. These six tools are:

1. Stakeholder identification and analysis—who is to be involved in the process
2. Sustainability appraisal—to identify and evaluate the economic, social and environmental impacts of the YSDAP actions
3. Sustainable development indicators for the monitoring process
4. Policy mapping—to identify relevant policies and their indications
5. Visioning—what future sustainable development in Yunnan might involve
6. Scenarios—a set of alternative futures.

The second phase involves putting the various tools into action via workshops and such procedures. YEDP is also providing guidance on the structure and content of the YSDAP.

The YSDAP will discuss various topics related to sustainability issues, but it is limited within the geographical boundaries of the province and does not include any consideration of transboundary impacts. According to the YEDP representatives, transboundary impacts have been excluded from the plan because of the very controversial nature of the issues.

Global initiative to study re-operation possibilities with the large dams

The Natural Heritage Institute (NHI) has recently launched a project called 'Global Survey of the Opportunities to Restore Aquatic Ecosystems Through Reoperation of Hydraulic Infrastructure' (NHI 2004). The purpose is to create a global survey and inventory of the most promising and feasible opportunities to modify the operations of major water storage and diversion projects to restore lost ecological functions in the downstream river basin. Additionally, local NGOs will be trained in the use of the tools and techniques created within the project. The background of the project involves the notion that due to narrow target setting, many large hydraulic projects have caused pervasive damage to freshwater, riparian and estuarine ecosystems.

The project aims at managing water projects to restore ecological values without jeopardizing their original economic objectives. The basic structure is to identify the restoration potential of selected dam schemes and thereafter provide a list of opportunities to the relevant national governments and international development assistance agencies. The purpose is to modify the operation of the water facilities to restore a more natural flow regime in the downstream channel and floodplain without substantially disrupting the irrigation, power generation, flood control and navigation benefits. The NHI plans to evaluate eighteen large-scale dam projects worldwide, selected according to specified criteria.

Initially, the project focuses on studying the already existing dams. Whether the evaluation could include projects that are still at the planning stage is still under consideration. Including them would allow ecologically friendly operation options right from the beginning. The upper Mekong dams have been suggested as one case study (Thomas 2004). They would form an interesting possibility in terms of considering the impacts of a cascade dam construction on an international river. It seems, however, that due to the data availability it is difficult to include the cascade in the NHI project.

Water policies

The highest regulative contribution in the Chinese water sector comes from the newly revised Water Law of 2002. The 1988 Water Law was renewed simultaneously with the institutional reform of the water sector. The reform included shifting most of those administrative water functions, which were earlier handled by other ministries, to the Ministry of Water Resources (MWR). The MWR works directly under the State Council and is now in charge of the unified administration and supervision of water resources throughout the country.

According to the Water Law 2002, river basin management agencies are supposed to exercise water resources management and supervision on state-designated major rivers and lakes. These agencies are set up by the MWR and include representatives from the relevant provinces. The state-designated major rivers include the Yellow, Yangtze, Pearl, Haihe, Huaihe, Songliao and Taihu rivers. According to the Water Law, comprehensive plans for other river basins that cross provinces are formulated by the same river basin management agencies jointly with the relevant departments of the provincial governments, and are submitted to the central level water administration after being reviewed and commented on by the provincial governments. In other words, the comprehensive planning of all trans-provincial rivers within China should fall under the seven river basin management agencies. For example, the Yangtze River Commission is responsible for planning on the Mekong, and the Pearl River Commission on the Red River. The working arena of the basin management agencies is, however, limited to water resources management, which in China is defined not to include hydropower related issues.

In general, decisions over large water projects are made at the central level, regardless of whether the river is interprovincial, transprovincial or international. The definition for a large hydropower project in China is that its installed capacity is greater than 300 MW (fifty–300 MW is considered medium and < fifty MW small scale). The central water governance consists mainly of the MWR, but there are other institutions involved as well. Some of the water related issues are still handled under other ministries and institutions, including the Ministries of Agriculture (irrigation), Traffic (river course navigation), Forestry and Construction (municipal water supply), the State Energy Company (hydropower) and the State Environmental Protection Agency (water environment, river basin ecological environment). Additionally, the State Council and its bodies take part in the decision-making when it comes to major water projects. Moreover, the level of involvement by the provincial and local governments in the decision-making process over international rivers has not been very clear. As can be noted, water governance is still a rather complicated matter in China, in particular in the case of large projects and international rivers.

Reflections on transboundary water issues

What is happening on the Chinese side of the great transboundary rivers can be partly explained by the aspects discussed above. In light of the economic challenges faced by Yunnan, the dams and other river development could be justified from the Chinese point of view, assuming the hydropower would significantly benefit local people. Unfortunately, it seems not to be so. For example, with the Manwan hydropower site in operation, electricity prices near the dam remain high. The Manwan plant sells electricity to the Manwan Power Supply Station at 0.17 *yuan*/kWh and the latter provides power at 0.38 *yuan*/kWh to Manwan Town (He and Liu 2002). When transferred further on to the townships, the price increases to 0.60 *yuan*/kWh and finally, the peasant households have to pay an average of 0.8–0.9 *yuan*/kWh and some are even charged as much as 1.6–2.0 *yuan*/kWh. In comparison, the equivalent civil electricity price in Kunming is about 0.402 *yuan*/kWh. The given reason for the extremely expensive prices is the huge losses during transmission, caused by the highly scattered nature of settlements and inefficient management.

Some common benefits will be created by improved transportation and other infrastructure facilities likely to be built along with the power plants. Creating jobs has also been suggested as one reason to give the projects a green light (Ma unpublished). Still, it seems that the poverty alleviation impacts of large dam construction in Yunnan will remain limited. Noting all the aspects, it seems clear that somebody other than the local people will collect the main benefits produced by the dams. The power companies, which have the right to develop the rivers and operate the power grids, will reap the profits from selling and transferring the electricity. The provincial government will benefit in the form of taxes paid from the company profits. In the future, when the water fee system in China is stabilized, the government will probably also collect taxes for the use of the water resources. The tax income could support poverty alleviation and environmental conservation if specifically allocated for such purposes. However, it is unlikely that the money gained from energy business taxation would be allocated directly for poverty and environment issues.

One consequence of the limited resources is that several regions within Yunnan are still extremely poor in scientific studies. The location in remote and mountainous areas and the lack of transportation access causes high costs of data collection. For instance, the source of the Red River hasn't yet been confirmed. The upland of the Irrawaddy River and the Sino-Burma border remain very poorly studied. Even on the Mekong, the study of the upstream regions is still weak and a unified information system is lacking. In a poverty-stricken province with frequent environmental disasters, there is not enough financial or other capacity to support data gathering. In this regard the lack of resources and the magnitude of challenges caused by the physical characteristics of the region seem to reflect the transboundary issues. There is a lack of such studies that would allow considerations of the transboundary impacts and decision-making based on scientific facts.

The purpose of this chapter was to discuss the linkages between poverty, environmental challenges and transboundary water issues in Yunnan. The major aim was to investigate whether solving the problems within Yunnan requires such an effort that insufficient resources are left to consider transboundary impacts. However, the linkage between the issues is multi-dimensional. It is true that the people of Yunnan are poor and it requires a lot of capacity to improve their livelihoods, particularly if simultaneously trying to conserve the environment. Nevertheless, it seems that the major projects causing transboundary impacts will not serve this purpose. He and Liu argue that in actuality the upper Mekong dam cascade was proposed in light of the national electricity shortage, rather than local power demands (He and Liu 2002). According to them, electricity generated for instance by the Manwan dam is mainly serving power demands in the Guangdong Province on the east coast. The extremely high electricity prices near the dam have been explained by the scattered households and electricity losses during transmission. Transmitting power to individual households in small amounts certainly increases the unit price of electricity. But if the purpose of the dams were to support local development, as the developers claim, there should be ways to subsidize local energy prices. Now it seems that the real benefit from the dam construction is going to the more developed parts of China.

Final remarks

Yunnan province has so far played a major part as the representative of China in several transboundary issues. Particularly within the Greater Mekong Subregion program, Yunnan is the officially named member instead of China. Due to the fact that several large international rivers flow out from China in Yunnan, the province acts as a water tower for the downstream regions. For these reasons, it would seem justified to ask of it more responsibility for the transboundary impacts caused by the major water projects within its territory.

Conversely, Yunnan is still one of the least developed provinces in China with a large population living in poverty. Simultaneously, the rich biodiversity and significant natural resources should be protected from destruction caused by human activities. As discussed in this paper, it is sometimes very difficult to integrate economic development, including poverty alleviation, and environmental conservation. In the case of Yunnan, this seems to be one of the major problems and requires significant attention. Following all the development challenges, the capacity in the province is scarce and not many resources are left for other issues, such as transboundary impacts of water projects.

Nonetheless, in the previous discussion it was shown that the large water projects in Yunnan will not necessarily help the local people. The two major issues seem to be poles apart—the valuable work of several institutions fighting poverty in Yunnan, and the large-scale river development. Additionally, considering that according to Chinese law the management of large water projects is directly under the auspices of the central government, the central level cannot retreat from the responsibility by letting Yunnan alone deal with the transboundary issues. It still retains control over many of the actors involved, including the state power companies. These actors should be obligated by the central level to increase cooperation with the other stakeholders. One of the very essential steps would be to increase information sharing between the energy companies and both domestic and international stakeholders.

The other option would be to give Yunnan more responsibility, but also to assign it more resources to fulfill this responsibility. With limited resources the province cannot be expected to handle much more than the significant internal challenges it faces. Additionally, having more financial recourses specifically allocated for the purpose and access to the relevant information would allow comprehensive academic research on the water resources and projects, including their impacts on the environment and livelihoods. Perhaps more importantly, it would also allow scientific studies on how to mitigate the negative impacts without significantly reducing the benefits that the dams are expected to bring. There have been some advanced initiatives made in this regard, among others the Natural Heritage Institute project to globally study reoperation possibilities with the large dams. However, without access to the relevant hydrological and dam operation data, it is impossible to perform such studies.

The planned and ongoing projects for poverty reduction and environmental protection in Yunnan are not able to deal with the transboundary issues. For example, in the YSDAP project presented above, transboundary impacts were excluded because of their controversial nature. The work within such projects is focused on efforts to alleviate poverty and environmental destruction in Yunnan, and they mostly deal with issues that are not too sensitive in nature. It is understandable that due to the limited resources, relatively small projects will have to exclude overly complicated issues. Moreover, due to the difficult access to data, financial gaps and political constraints, the Chinese institutions lack the capability to extend their activities significantly across the borders. Only a mechanism with ongoing and significant support from the central government, access to relevant data, and adequate financial resources could really consider the transboundary dimension. Once launched, such a mechanism could be linked with the work done by the downstream institutions, such as the MRC.

On the Chinese side, some of the constraints could be removed by strengthening the existing scientific institutions and authorizing them more widely, and by launching regulations that would increase the public availability of data. As far as these constraints remain, it is unlikely that a method for managing international rivers on a river basin base will be found. For example, both the Chinese researchers and the MRC-affiliated institutions have carried out hydrological modeling about the impacts of the upper Mekong dam cascade. Nevertheless, both lack adequate data from the other side, and therefore the results are not able to cover the whole river basin. Consequently, many questions of the real impacts remain unanswered.

References

Asian Development Bank (ADB) (2004), *Poverty profile of the People's Republic of China*, Manila: Asian Development Bank, available online at *http://www.adb.org/Documents/Reports/Poverty_Profile_PRC/default.asp.*

ADB (1998a), *Yunnan Dachaoshan power transmission project in the People's Republic of China*, Summary environmental impact assessment, Manila: Asian Development Bank.

ADB (1998b), *Report and recommendation of the president to the board of directors on a proposed loan and technical assistance grant to the People's Republic of China for the Yunnan Dachaoshan power transmission project*, Manila: Asian Development Bank.

Asian International Rivers Center (AIRC) (2004), Web pages of the Asian International Rivers Center, available online at *http://www.lancang-mekong.org/index-e.asp*

Butler, T. (2004), 'Thai groups battle new China dam', *BBC News Online*, BBC.

Cheung, R. (2004), 'Beijing to approve Yunnan dams plan', *South China Morning Post*, November 2.

China Daily (2004a), 'Forum pushed bid to develop the west', November 19.

China Daily (2004b), 'Western region development to lead agenda', November 19.

China Daily (2004c), 'More bank loans to help to develop the West', November 19.

China Economic Review (2002), *The China Business Handbook 2002*, Hong Kong: Alain Charles Publishing.

Dore, J. and X. Yu (2004), *Yunnan hydropower expansion, update on China's energy industry reforms and the Nu, Lancang and Jinsha hydropower dams*, working paper, Chiang Mai University and Green Watershed.

He, D. (2004), Personal communication, Asian International Rivers Center, Yunnan University, Kunming.

He, D. and X. Liu (2002), 'Water resources management in Yunnan: Fragmented institutional structures and resulting conflicts', unpublished manuscript.

He, D., and X. Liu (2001), 'Equitable utilisation and effective protection of sharing transboundary water resources in international rivers of western China', *Journal of Geographical Sciences,* 11 (4), October: 490–500.

He, D., C. Liu and Z. Yang (2001), 'The study on the sustainability of the international rivers in China', in D. He, C. Zhang and K. Hsiang-te (eds), *Towards Cooperative Utilization and Co-ordinated Management of*

International Rivers, Tokyo: The United Nations University and New York: Science Press.

He, D., W. Zhao and Y. Feng, (2004), 'Research progress of international rivers in China', *Journal of Geographical Sciences*, supplement 14: 21–28.

He, D., W. Zhao, Y. Feng, X. Liu, S. Gan, J. Gou and M. Zhang (2002), *Water Resources Policy Review and Institutional Capacity Analysis*, Yunnan Environment Development Project, Research report No.2001D0002Z, Asian International Rivers Centre, Yunnan University.

He, D., W. Zhao and J. Gou (2003), 'Study on the reasonable utilization of transboundary water resources and environmental protection in southwest China', in Proceedings of the first international symposium on Southeast Asian water environment, biodiversity and water environment, October 23–25, Bangkok: Asian Institute of Technology.

Heilig, G. K. (1999), *Can China Feed Itself? A System of Evaluation of Policy Options*, CD-ROM, Laxenburg: International Institute for Applied System Analysis (IIASA).

IRN (2004), Web site of the International Rivers Network, available online at *http://www.irn.org*

IRN and Tibet Justice Center, (2004), *Media Advisory: Chinese Premier suspends plans to dam Nu/Salween*, April 7.

Kummu, M., J. Sarkkula and O. Varis (2004), 'Sedimentation and Mekong upstream development: Impacts on the Lower Mekong Basin', Shanghai: IAG Yangtze Fluvial Conference, June 24–27.

Ma, H. (2006), *The sustainable development of Lancang River region: A 21st century challenge of the balance between development and reservation*, JCAS Symposium Series, No. 25.

Makkonen, K. (2003), *China's role in IWRM in South and Southeast Asia*, Espoo: Helsinki University of Technology.

McDonald, K. (2008), 'Hydropower on the Nu: One river, many perspectives', *International Rivers* March 19, at *http://www.internationalrivers.org/en/china/nu-salween-river/hydropower-nu-one-river-many-perspectives*.

NBS (2002), *Poverty Monitoring Report of Rural China 2002*, China Statistical Yearbook 2002, National Bureau of Statistics, Beijing: China Statistics Press.

NHI (2004), 'Global survey of the opportunities to restore aquatic ecosystems through reoperation of hydraulic infrastructure', project description, Berkeley: Natural Heritage Institute.

New York Times (2004a), 'In a surprise move, Wen cites concern for environment', April 8.

New York Times (2004b), 'Dam project threatens Chinese "Grand Canyon"', March 10.

Office of the Leading Group for Western Region Development of the State Council (2004), Web site available online at *http://www.chinawest.gov.cn*

Piyasvasti, A. (1999), 'Development of a regional energy market in the Mekong area', presentation, Energy Week, 7 April 1999, Washington D.C.

Plinston, D. and D. He (2000), *Water Resources and Hydropower*, Policies and Strategies for Sustainable Development of the Lancang River Basin, Sector Studies for ADB TA No. 3139-PRC, Landcare Research New Zealand Ltd.

Reuters (2004), 'China in for extreme weather as climate change felt, report says', Reuters report November 9.

Sarkkula, J., M. Kiirikki, J. Koponen and M. Kummu (2003), Ecosystem processes of the Tonle Sap Lake, Ecotone II - 1 workshop, Phnom Penh/ Siem Reap, Cambodia.

Sarkkula, J., E. Baran, P. Chheng, M. Keskinen, J. Koponen and M. Kummu (2004), 'Tonle Sap Pulsing System and Fisheries Productivity', SIL XXIX International Congress of Limnology, Lahti, Finland.

Thomas, G. (2004), Personal communication, Berkeley: Natural Heritage Institute.

UNESCO-IHE (2004), Web pages of the UNESCO-IHE 2004 Institute for Water Education, available online at *http://www.unesco-ihe.org/Project-activities/Project-database/Integrated-Ecosystem-and-Water-Resources-Management-of-the-Upper-Mekong-River-Basin*.

US Embassy Beijing (2000), *Dianchi: Anatomy of a Polluted Chinese Lake*, report from U.S. Embassy Beijing, April 2000, available online at *http://beijing.usembassy-china.org.cn/report0400water.html*

World Bank (2002), *For $1 a day, Country Assistance Strategy for PRC*, Washington DC: World Bank.

WUP-FIN (2003), 'Modelling Tonle Sap for environmental impact assessment and management support', MRCS / WUP-FIN Project, Final Report, Mekong River Commission, Phnom Penh.

Yang, H. Z. (2001), 'Strategic focus of eco-environment construction and key regional harnessing' (in Chinese), in Conference Symposium of Effective and Sustainable Management and Optimum Allocation of Resources in the West, Chengdu: Sichuan Science and Technology Press: 114–119.

Yunnan Daily (2003a), 'Liuku hydroelectric station construction ready to start', July 19.

Yunnan Daily (2003b), 'Hydropower Planning Report of Middle and Lower Reaches of Nu River discussed in Beijing', August 15.

13
Environmental Changes of Handgarte Pastures of Altai over Fifty Years: A Case Analysis of Xinjiang Altai Grassland

Tsui Yenhu

Environment problems or issues are a global phenomenon evident in almost every corner of the earth, though their forms and impacts are uneven due to varying ecological, societal and developmental conditions. Since environmental problems are a global reality, those that occur in one area impact on adjacent areas, even on distant regions, and international concern and cooperation is essential for human beings to overcome these ecological challenges. Dialogue and cooperative studies on specific environmental issues involving specialists from different disciplines and stakeholder countries will be able to make special contributions to the issues at hand.

This chapter argues that while it is important to understand China's environmental problems at a national and regional level, it may be more important to examine these issues from grassroots perspectives. This is perhaps controversial, but particularly important when understanding what is happening in the environments of local people where their activities, whether production or lifestyle related, exert impacts of various kinds on the eco-environment in which they live. It is with this assumption that the

Photo 13.1. Migrating Kazak herders, Handgarte

present chapter reports a case study of environmental changes over the past fifty years of Handgarte Township (*Xiang*), Altai Prefecture, a small pastoral area of the Altai Steppe of North Xinjiang, China.

The data used in the chapter mainly come from fieldwork carried out by the author and his colleagues over the past ten years. Further data is drawn from the indigenous knowledge of the local people, Uranghai Mongol and Kazakh herders, on the environmental changes of the area. This knowledge is of particular importance, for the perceptions of local peoples of changes to their environment may paint a picture closer to reality than is otherwise possible to ascertain.

The case study site: Handgarte Mongolian Xiang

Handgarte Mongolian Xiang is located in a low undulating part of the Altai Mountains, forty-seven km as the crow flies from Altai City, the capital of Altai Prefecture. The Xiang shares a border with Mongolia Republic in the east. The total area is about 1270 km^2, which can be divided into two parts: the hilly area with Handgarte River flowing through it and the valley of Erqis River. Most of the area is at approximately 1000 m above sea level but a small part located in the river valley area is at 540 m above sea level. Geographically, this area is characteristic of high mountain ranges in the north and low hilly areas and river valley plains in the south. The center of Handgarte Xiang is a village, also named Handgarte, but the Han Chinese of Altai used to call it Lamazhao as there was a large compound of lama temples which were destroyed in the 1940s during a Kazakh uprising.

The major ethnic groups of Handgarte are Uranghai Mongols, Kazakhs and Han Chinese, whose combined population comprises about 96.8% of the total. Other ethnic groups represented are the Uygur, Hui and Tatar. The population was 4588 in 2002, of which the number of Uranghai Mongols was 1396, Kazakh 1761, and Han Chinese 1384. The number of Uygur, Hui and Tatar was just 147. According to local memories, Uranghai Mongols are believed to be the first inhabitants of the area, with Kazakhs arriving in the late nineteenth century from Russian Kazakhstan. A small number of Han Chinese came to settle in the area in the nineteenth century but most arrived in the mid 1960s.

The most important resources of Handgarte are pastures, forests, water and minerals. Of the total territory of the Xiang, about seventy-four percent is pastures which lie in river valleys, hilly areas and high mountains. Over the past fifty years, there has been an increase in the area of farmland mostly located around the villages that have developed since the 1960s.

The subsistence patterns of the Handgarte people vary according to ethnic groups. More than seventy percent of the Kazakhs and more than sixty percent of the Uruanghai Mongols are engaged in nomadic pastoralism while all of the Han Chinese, Hui, Uygurs and Tatars are engaged in farming, growing grains and legumes (wheat, maize and beans) and raising animals in pens. In the past fifty years, a small but increasing number of Kazakhs and Uranghai Mongols have settled down and taken up some farming.

The difference in subsistence patterns determines the patterns of resource use and residence as well as administrative systems. Farmers mainly live in the five farming villages of Hoblet, Arqiat, Tarat, Qiaorhi and Dablehart, where their houses lie in dense rows. Nomadic pastoralists are administratively organized into two separate villages called Josat and Kokmodun (the nomadic herders call them '*muye yidui*' and '*muye erdui*', meaning 'Number one pastoral village' and 'Number two pastoral village'). They have fixed houses only in the winter pastureland, where they reside only in wintertime (from early November to mid-March), and their houses lie far apart. In each of the villages there is a village council that performs the administrative functions to some extent. However, the functions differ respectively in pastoral villages and farming villages due to the differences in subsistence patterns. For example, the major administrative functions of the pastoral village council are to organize seasonal migration and to provide nomadic herders with services, such as settling disputes between herders and farmers, preventing and controlling epidemic diseases in animals and collecting taxes according to the laws and regulations of the local government. Alternatively, the councils of the farming villages must make plans for farming and provide the farmers with services, such as introducing new farming technologies, distributing chemical fertilizers and water resources for irrigation and helping farmers in selling their grain and other produce.

In fact, to a certain extent, the existence of two kinds of villages is a result of the changes in subsistence and environment over the past fifty years. Before 1950, there was almost no farming population and settlement in Handgarte except some separate houses of Russians who arrived at the beginning of the twentieth century and a few Han Chinese peasants whose ancestors came here with the troops of the Qing dynasty about two hundred years ago.

Changes in the environment of Handgarte

In this section, a description of the environmental changes is presented and some specific data are offered in order to illustrate alterations to the environment of Handgarte and their causes. The author's conclusions, which

Photo 13.2. Lama Temple and the Lama, Handgarte

might be considered subjective, are not the primary focus. Rather the aim is to reveal what has been happening to the environment that remaps the land in terms of its ecosystem and life patterns of local people. It is believed that ten years of fieldwork in Handgarte, beginning in 1993, has helped the author to understand in a gradual and systemic way the environmental changes in this pastoral area. This has afforded the author a perspective on the changes that extends beyond the timeframe of the fieldwork and has also enabled the author to observe the process through which the changes are continuingly occurring.

In the past fifty years, the changes in the environment of Handgarte have taken place in three correlative aspects: deforestation, degradation of pastures and water pollution. All these changes have resulted in a dramatic alteration to the ecosystem in the area and in turn, they have seriously influenced the living conditions of the local people.

Deforestation

Handgarte lies in an area extending from the southern part of Uhili Mountain of the Altai Mountain Ranges to Erqis River Valley. The mountain itself used to be covered with old-growth forests; the coniferous forest growing above the area at 1300 m above sea level and the broadleaf forest of Altai composed of Altai birch trees, poplar trees and various bushes in the area from that altitude down to the basin below. The forests used to provide Uranghai Mongols with rich resources of wild animals. Hunting was thus the dominant subsistence form before 1949. For the Kazakh nomadic herders, the forests were not only the source from which they obtained firewood, wood for yurt-frames and covered pens, but also pastures for their animals. Almost all the elderly people we interviewed told a similar story of the state of the forests prior to 1960, saying: 'our animals could not get into the woods because they were very dense', and 'we could get firewood for the whole season in two days when we stayed in the woods'.

The past fifty years has seen the gradual disappearance of the forest in most areas except in high mountains where the forests become less dense. In some areas there are no forestlands at all.

Case 1: Disappearance of the bush woods in the hilly area

In the hilly area between the Uxili Mountain and the Handgarte Basin there used to be very dense woods composed of birch, poplar and various kinds of bushes which had extremely thick coverage along the hills and slopes. This woodland is the main location where herders gather firewood and branches for their yurts and folds. Before 1957, the local elderly herders told us they had often lost their animals there because of the density of the trees. Now there are no woods at all in this area except for some very sparse trees and bushes that can be seen in some parts far from the villages. An old Uranghai herder told us that in the 1950s, it was very difficult for his horse to go through the woods and during winter he could hunt many wild animals: 'There were so many that you did not go too far that you saw them. Where are they now?' The decisions made by the local governments and an increasing settled population are two major factors that caused the changes to the woods.

In 1958–59, hundreds of people, local or from Altai City gathered in the Handgarte Basin to make 'iron and steel' as there were rich mineral resources and fuel in the area. Not much steel and iron was produced, the only consequence was the felling of thousands of trees and bushes, which extensively damaged the woods in the Basin and the undulating slopes around it.

Since 1960, with a rapid population increase, especially of farmers, the woods were damaged even more severely and fatally because of the new farmers' need for firewood. Duan is a clerk of the local post office and he was born and raised at Handgarte. He said that when he was a small child, one of his major tasks was to collect firewood for his family, and almost every day in summer and autumn he had to gather about twenty-five kilograms of firewood. Five of his family members had to collect firewood for the winter season, causing the disappearance of about twenty-five m^3 of wood, which led to the damage of woods of about three to four *mu* (about 0.32 ha.).

In addition to the people of Handgarte, people from the neighboring areas also collected firewood from the same forestlands. Some elderly farmers told us that because there were not many trees in the Basin and surrounding hilly slopes, they had to go up to the mountain forest to gather firewood. The woods in the Basin and hilly slopes finally vanished in the late 1970s, and people had to go far up into the mountains to fell big trees for firewood.

It is the same story with the woods in the valley of Handgarte River. Before the 1960s there were dense woods in the river valley from the upper area to the low reaches. 'The trees were so dense that it was hard to go through if we wanted to go to the water', said Jinser, a Uranghai man who was the head of the Grassland Monitoring Station of the Xiang in 1993, 'but now where were they (trees)?!' According to him, 'At least, the forest line (of Handgarte) has shrunken up to the high mountains for about forty km in the past thirty years'.

The deforestation resulted in some serous environmental problems in the area. For example, in the area of Tiemur Village which is situated in a mountain valley, before the 1960s, the forest covered almost seventy percent of the land but now this coverage has dropped to no more than fifteen percent. Soil erosion in that area is the most serious in the Handagrte region. In the bottom of the valley, there are many deep ravines eroded by water rushing down from the mountains, some are as deep as three to five m. The village of Tiemur is now the poorest in Handagarte, and sixty percent of the villagers have to live on relief grain and money from the government.

Degradation of pastures

The territory of Handgarte is a part of the Altai Steppe, and pastoralism used to be the major form of subsistence system of the local people. Almost seventy percent of the land is pasture, the most important resource for the Uranghai and Kazakh people. However, the most obvious changes to the environment have happened to the pastures and these alterations have made it impossible for the nomadic population to subsist as they once did, as the pastures have deteriorated so badly that they can no longer support the increasingly large herds of animals. Of the causes of the pasture degradation, three are believed to be significant: reclamation, over-carrying and grazing, and mining.

In the 1950s, agriculture played a very small part in local subsistence. According to the memories of the elderly people and local archives, the total area of farmland was no more than 150 *mu* (about ten ha). Now the total area of farmland is more than 40,000 *mu* (about 2667 ha) and the reclamation of grassland has not yet ceased.

There have been three phases of grassland reclamation in the Handgarte area. The first phase occurred in the early 1960s, when Altai County needed more grain to support the increasing population both in the town and in rural areas. Pastures of about 30,000 *mu* (about 2000 ha) were reclaimed at that time.

Photo 13.3. Location of Handgarte in the Map of Altai Prefecture, Xinjiang

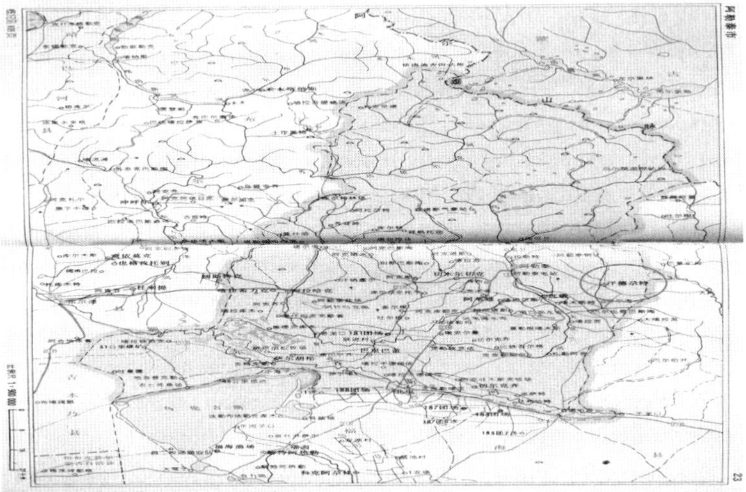

Photo 13.4. Spring pasture in 1999, Handgarte

Case 2

The Handgarte Basin is a small one with an area of about forty-six square km. The pasture there used to be the best in the area as the Basin is flat with a river flowing through it. In 1968, in order to implement the decision of the Central Government to 'Take Dazhai (a famous village in Shanxi Province) as the model for agricultural development', a section of the pasture of the Basin, about 10,000 *mu* (about 650 ha) in area, was reclaimed. In the following three years, pasture about 2800 *mu* (about 186.6 ha) was reclaimed, and every farming family and newly settled herding family that lived near Handgarte Village, the Xiang Government Site, had farmland of about twenty *mu*. However, due to shortage of water for irrigation, lack of farming technologies and poor management, the reclaimed land was deserted three years later because of poor produce.

In 1983, the Altai County government decided to restore the pasture in the Basin and an airplane was used to spread seeds. The local people hoped that the seeds could be spread in a uniform way so that the reclaimed pasture could be restored and grow well. In order to stop animals from going into the restored pasture, a wire lead was built around it. The following year Handgarte Village decided that this piece of pasture should not be distributed to herder families in the way that most of the pastures of the Xiang had been distributed, based on a new land-use system—the responsibility system that was implemented in rural China in 1984. The village hoped that the restored pasture could be used as common land when it was regenerated to the extent that the varieties of grass grew and the coverage regained what it was before it had been claimed. Eight years later, to the disappointment of the villagers, the coverage did not reach the degree that they had hoped, but people were still patient and expected to see the day when the pasture finally became what it had been twenty years before.

However, in order to offer pastures and farmland to those newly established families, in 1992, the Xiang Government made two decisions. The first was that a part of the pasture should be distributed to some of the families, and the second was that on the southern edge of the pasture, newly established farmer families could claim a certain area to grow their crops.

This leads to a new type of pasture degradation, as each of the newly established families obtained too small an area which could not support their animals. Overgrazing was thus a common practice and the quality of the pasture was again deteriorated in terms of coverage and height.

In 1999, this pasture appeared to return to its condition prior to restoration—many patches became bare of grass-cover and where grasses appeared coverage was very thin. More land was reclaimed and on the eastern edge of the pasture, a new iron ore mine was opened and mineral waste became scattered here and there. The restored pasture in the Basin experienced a second round of degradation.

Case 3

In 1988, some farmers began to build a clay fence separating the pastures from their farmland, about 800 m in length and 0.5 m in width. This fence aimed to determine the boundary between the sedentary population and their subsistence system and the nomadic herding population and their livelihood. In this manner farmland and crops could be protected from grazing animals. This clay fence continued to be extended in the following years. In 1999

Photo 13.5. Early Summer pasture in 1993, Handgarte

when the author conducted fieldwork the fence was many times longer than it had been in 1993, about seventeen km in length. Almost all the farmers who shared the boundaries between their farmland and the pastures of the nomadic herders joined the farmers who had started building the fence in 1988. This event is particularly interesting as this is what had occurred more than two thousand years ago in North China when the sedentary people built the Great Wall to stop nomads from entering their territory. This wall of Handgarte, though much smaller and shorter than the Great Wall, suggests that the conflicts over territory and resources between nomads and sedentary people of ancient China have persisted over thousands of years.

In fact, conflicts over resources between herders and farmers in Handgarte have never ceased; they have in fact become increasingly serious because of the reclamation of grassland and expansion of farmland. Farmers complain that herders' animals come to their farmland and damage their crops while herders complain that reclamation of more and more pastures narrows their 'territory' so greatly that it is very hard for their animals to have enough space to move about. Although the clay wall cannot compare to the Great Wall in scale and significance, the motives of the farmers are surprisingly identical to those of the sedentary population of ancient times, that is to say, the physical boundaries between nomadic people and the sedentary people in terms of their respective subsistence and different strategies to access resources lead to psychological boundaries evident in conflict resolution methods—building walls to protect their right to resources. A new implication in the Handgarte case is that the environmental crises is due to the expansion of sedentary subsistence systems in an area where traditional nomadic pastoral subsistence systems were dominant and had been practiced for thousands of years.

What has happened in the Handgarte Basin has actually occurred in other parts of the area. For instance, in 1993 there was no farmland in the area close to that part of Handgarte River near the village where pastures were good, though the valley forest had vanished. In 1999, patches of farming fields extended to the waterline of the river, and these patches were fenced with dry tree branches or bush stems. Herders said that they were not able to move their animals to the pastures along the river nor were they able to move their animals to drink water in the river. They also complained that more and more traces of chemical fertilizers and even agricultural chemicals penetrated into the river from the farm fields as farmers became increasingly reliant upon them. 'Horses would not like to go to the river to drink water, for they are very sensitive to unnatural smells in it', a Kazakh herder told us.

It is not only the local population that is involved in the process in which pastures are claimed and deteriorated in Handgarte, but also people from other areas, including those from other parts of the country. Since the latter 1980s, some of these immigrants have been been engaged in digging medicinal herbs such as licorice and ephedrae roots in the Handgarte grassland. Their activities naturally do great damage to the grassland. For instance, only in 1993–94, the area of pastures that was extremely damaged by such activities amounted to 40,000 *mu* (about 2560 ha).

The results of the surveys regarding changes in the pastures of Handgarte in 1993 and 1999 show that some obvious negative consequences resulted from the changes, mainly in the coverage and length of the grasses, and the productivity of pastures. The coverage per square meter reduced from about sixty to eighty percent to twenty-eight to forty-four percent and the height of grasses dropped from twenty-five to thirty cm to five to eight cm. The productivity of grassland has reduced by about fifty percent. What is worse is the disappearance of fine quality grasses and the increase of amount of grasses that animals never touch. This is actually what has been happening to most grassland areas in Xinjiang over the past fifty years.

Environmental changes in Handgarte River Valley

Handgarte River, about fifty km in length, rushes down from the Tiemur-baz Snow Mountain and passes through a mountain valley, rushes into the Five-finger Hill Basin encircled by rocky hills, and flows down to the Handagarte Basin. The river then flows into a low hilly valley, finally merging into one of the major braches of Erqis River: Kolan River.

It is a small river with a very limited runoff. However, it is a lifeline for the farmers whose farmlands lie along the river or in the Basin, and for the nomadic herders whose spring-autumn pastures are along the hilly areas

Photo 13.6. Degraded pasture in 2002, Handgarte

on the banks of the river. Every year nomadic herders move their animals from either the winter pastures in the Erqis River Valley in early March or from their summer pastures in the high mountains in early October to this part of Handgarte River valley where they remain for more than five months. The conditions of the ecosystem of Handgarte River are vital for the livelihoods and animal husbandry of the people there. Therefore, changes to the ecosystem of the river results in environmental changes to the valley area, which in turn have substantially significant impacts on the existence of the local people.

In fact, of the environmental changes that have occurred in the Handgarte area in the past fifty years, the most dramatic is changes to the river ecosystem. These changes are an indicator of the eco-environment quality of the area.

Of the eco-environmental changes that have happened in the river and its valley, the most obvious changes have occurred in the river valley woods in the middle and lower reaches. The old-growth woods composed of birch, poplar, willow and various kinds of bushes are now almost disappeared, except in some parts where there are one or two lonely trees. Although the woods still look intact in the upper reaches, the number of cases of illegal felling has increased year by year. Many trees were cut down by either farmers or those who came in search of gold. The disappearance of the woods in the middle-lower reaches of the river dramatically 'remaps' the eco-system, attested to by the facts that the water is becoming muddy with more and more siltation as a result of erosion and the runoff is decreasing.

Another change to the river ecosystem is the gradual disappearance of some adjoining wetlands. In 1993, there were a number of wetland areas both along the river and in the basin area not far from the river. Small or large, wetlands perform an important function in maintaining the micro-ecosystem of the river and pastures. In these areas the author sighted many birds and varieties of grasses, the quality of the latter being the best in comparison with other areas. However, in 1999, what we saw was dried-up and bare land. In the dried wetlands, almost no grass and birds were seen.

Yet another environmental change in the river is evident as a large area of the woodlands and pastures was reclaimed and more and more chemical fertilizers and agricultural chemicals are used with the runoff penetrating the soil and flowing into the water. The extinction of fish in the river has been the result. Before the 1980s, there were several species of fish in the river with some rare and indigenous only to Altai Steppe area. Jinser, a key informant, told us that when he was young, he often bathed in the river and at that time 'you could catch them (fish) as easily as you catch a goat!' Another Uranghai Mongolian told us that in the 1970s he and other male villagers could 'catch fish in the river by hitting them with sticks'. 'What has happened to them?' Jinser asked us with a puzzled look on his face. Actually, the river fish were killed by chemicals, but the local people do not realize that their use of agricultural chemicals is the main cause of the extinction of fish species in the river.

In conclusion, in the past fifty years, the environment of the Handgarte area has changed to an extent that negative consequences are seen in the forests, grasslands and waterways. These changes in turn have brought about negative impacts on the living conditions and development of the community. The annual income of families, including herder and farming families, increases slowly, and there has been a visible decline in the annual income of some herder families due to the degradation of their pastures in recent years.

The causes of environmental change in Handgarte

It seems paradoxically ambiguous to see the obvious negative changes of the environment in Handgarte, for it seems that this area does not have some of the visible factors that cause environmental problems in most areas of China. For example, serious problems that occur in most other areas such as over-population and a lack of natural resources are not found in this area. The average number of people per square kilometer in Handgarte is 3.54, which is much lower than that of most parts of China. Handgarte further possesses a plethora of natural resources, such as pastures, forest, water and mineral resources. What, then, has essentially caused the environmental changes? This is a question to which we have been trying for almost ten years to find answers.

In order to resolve this matter, we have to look into some basic factors that might be correlative with the environmental changes in this area.

Population growth

Although it appears that there is not a serious over-population problem in Handgarte, there *is* a population problem related to the problems we are now discussing. Our surveys prove that population growth in Handgarte has increasingly exerted pressure on the environment in relation to resource-use and subsistence patterns. In the early 1950s, the population was no more than 1300, and resource-use and subsistence patterns were that of nomadic pastoralism characterized by using grassland resources, such as pastures, water and forest, in a cyclical way that enabled the natural ecosystem to have enough time and space to recover. After the mid 1960s, traditional farming gradually became another important subsistence form and consequently, densely inhabited settlements appeared as evident in the five villages of Hoblet, Arqiat, Tarat, Qiaorhi and Dablehart. This changed the traditional resource-use pattern and at the same time pressures on the environment increased. This is why the woods of the low hilly areas disappeared in this period. In the 1990s, the population number was four times greater than that of the 1950s and it was the demands of this growing population for more resources that led to the increasing demand for farmland which could only be satisfied by land reclamation. This is why there has been an unceasing reclamation of pastures in the Handgarte Basin. Rapid growth of the pastoralist population also required a large number of stock to provide subsistence means, which in turn need more pastures. But the area of pastures is limited and this is why overgrazing and over-carrying occurred—the main cause of the degradation of pastures.

Expansion of traditional agriculture

It is noted that the environmental changes of Handgarte have been accompanied by increasing traditional agricultural practices, growing crops in particular. The rapid development of the area changed the subsistence pattern of the local population. In the later 1960s, both nomadic pastoralism and traditional agriculture were the major subsistence forms, and in the 1970s to 1980s traditional farming became the major subsistence means for sixty-five percent of the population. In the 1990s traditional agriculture replaced the position of nomadic pastoralism as the dominant subsistence means with more and more nomadic families settling down and becoming 'sedentary farmers'.

It is incorrect to say that agricultural development directly caused the environmental problems in general, and neither is it accurate to assume that traditional farming practices are the main cause of the problems. What is at issue here is whether the practice of traditional farming is the only way for the development of an area like Handgarte, as is believed by the local governments and people, and whether the geographical and ecological conditions of the area are suitable for large-scale traditional agricultural development.

Handgarte is situated in an area in which most of the land is either low hills or high mountains with very thin soil cover (about ten to fifteen cm thick). The annual rainfall is about 340–400 mm and the annual temperature is between -27°C– +23°C. This area, like all other areas of the Altai Steppe, historically had nomadic pastoralism as the main subsistence form of the local population, and farming production had never gained the position it has until now. The intact nature of grassland in terms of space and coverage is the major factor for ecosystem stability in the area. The environmental history of grassland in many other regions suggests that the eco-expansion of cultivation into grassland may cause serious environmental problems and even disasters, as was the case in the Kazakh Grassland of former-Soviet times in the 1960s and the North American Plains in the 1930s. As to whether this suggests that traditional agriculture is not a beneficial way to manage the ecosystem of the grassland economy remains to be explored. The Handgarte case at least proves that the over-development of cultivation partially results in negative changes of the eco-environment as it causes deforestation, pasture deterioration and water pollution.

Policy-orientation

It is also found that policy-orientation, especially non-sustainable policy-orientation, of the local governments is an important factor that causes environmental changes in the area. The reclamation of Handgarte Basin in the mid-1960s serves as a good example for this point. When local governments make non-eco-friendly policies on resource- and land-use of the grassland, the results may be disastrous for the environment. In recent years, the government of Handgarte Xiang decided that cultivation should be the key strategy for economic enhancement. This will certainly result in more pasture reclamation not only in the Basin but also in the pastures in hilly areas. The number of the herder families should be strictly controlled or should not be increased and newly established families should turn to farming as their subsistence form. It is clear that these policies will bring about more pasture reclamation in order to satisfy the needs of the increasing sedentary population.

The causes of environmental changes in a given area are complex, as is the case with the situation of the Handgarte grassland. However, it is believed that the above three factors are the most important.

Conclusion

In the past fifty years, the grassland of Altai Steppe area as well as that of Xinjiang has been undergoing dramatic changes which are basically deforestation, pasture deterioration and water pollution. According to official sources, of the total area of the grassland (1.2 billion *mu* in area, about 75,000,000 ha), eighty-five percent have been degraded and the area of seriously degraded grassland is more than forty-six percent of the total. Deterioration of grassland has brought about serious environmental consequences, desertification being the most striking, where the desert area is increasing by 150–180 km^2 every year. The causes for this change are complex, but it is not natural factors but social, institutional and cultural factors that are the most important.

The Handgarte case suggests that non-sustainable development strategies for pastoralist areas pose the major danger to eco-environment stability, and this analysis proves that eco-expansion of cultivation and cultural expansion of traditional agriculture into pastoralist areas like the Altai Steppe are two key aspects of non-sustainable development strategies.

Recent policies issued by the Chinese government for returning reclaimed farmland to pastureland and forestland may be understood as the great efforts to reverse the worsening environmental situation in the grasslands of China, and these policies have brought certain positive changes. However, the implementation of these policies is dependent on changes in the current policy-orientation and decision-making processes of local governments which in many areas still encourage the reclamation of pastureland. It will take desperate measures to pursue truly sustainable development strategies in the grassland areas of China.

References

Gregory, C. A. and J. C. Altman (1989), *Observing the Economy*, London: Rutledge.

Hann, C. M. (ed.) (1994), *When History Accelerates: Essays on Rapid Social Change, Complexity and Creativity*, London: The Athlone Press.

Hansen, A. and A. Oliver-Smith (1981), *Involuntary Migration and Resettlement: The Problems and Responses of Dislocated People*, Boulder: Westview Press.

Humphrey, C. and D. Sneath (eds) (1997), *Culture and Environment in Inner Asia*, Volumes 1 and 2, Cambridge: The White Horse Press.

Humphrey, C. and D. Sneath (1996), *Pastoralism and Institutional Change in Inner Asia: Comparative Perspectives From The MECCIA Research Project*, London: ODI, Pastoral Development Network.

Leybourne, M., R. Jaubert and R. N. Tutwiler (1993), *Changes in Migration And Feeding Patterns Among Semi-Nomaic Pastoralists in Northern Syria*, London: ODI, Pastoral Development Network.

Ma, X. (ed.) (1992), *The Annals of Altai City* (in Chinese), unpublished manuscript.

Mirzkhan, J. (1999), *History and Folklore of Kazaks* (in Chinese), Urumqi: Xinjiang People's Publication.

Moran, E. F. (2000), *Human Adaptability: An Introduction to Ecological Anthropology*, second edition, Boulder: Westview Press.

Plattner, S. (ed.) (1994), *Economic Anthropology*, Stanford: Stanford University Press.

Policy Research Section of CCP Xinjiang Committee (1988), *Pastoral Society of Xinjiang* (in Chinese), Guangzhou: Agriculture Publications.

Sinor, D. (ed.) (1990), *The Cambridge History of Early Inner History*, Cambridge: Cambridge University Press.

Su, B. (1989), *A History of Kazak Culture* (in Chinese), Urumqi: Xinjiang University Press.

Tsui, Y. (2000), 'A study of sustainable development of micro society of Northwest China' (in Chinese), *Journal of Xinjiang University*, 3: 37–42.

Wang, Z. (1998), *A History of Animal Husbandry Economy of Ancient Xinjiang* (in Chinese), Urumqi: Xinjiang People's Publication.

Xue, Z. (ed.) (1998) *History of Social life of Ancient Xinjiang* (in Chinese), Urumqi: Xinjiang People's Publication.

Yang, T. (1991), *On Migrating Herding* (in Chinese), unpublished manuscript, Urumqi: Xinjiang Academy of Social Sciences.

14
Eco-historical Background and the Modern Process of Desertification in Mu-us Sandy Land with Reference to Pastoral Life

Kobayashi Tatsuaki and Yang Jie

Ecological structure as the background of desertification

Mu-us Sandy Land is located on the ecotone between the humid and arid temperate zones of East-Central Asia (Figure 14.1). This area is a temperate grassland region of biome. The annual mean temperature is 6.4 degrees Celsius and the annual precipitation is 360 mm. The total area is 40,000 km^2. Mu-us Sandy Land spreads out on the Ordos Plateau with an altitude of 1000 m, surrounded by the Great Bend of the Yellow River. The eastern and southern borders are adjacent to the loess ('yellow earth') plateau, and the western end is delimited by the Ningxia Plain. Though the total percentage of sand dune is as high as thirty-six percent, it is not called a 'desert' but instead a 'sandy land', due to significant cover of grassland and scrub.

The formation of a sandy land is influenced by climate. Here, summer rain falls as fronts cross the lands. If the continental anticyclone retains strength, the front seldom reaches the north from the southeastern humid region even in summer. In such years, temperatures remain low and sunny days continue during summer. It is thought that during the Ice Age this area was dominated by an arid climate. Under an arid climate, desert covered this area. Sand dunes were active all over this region. Plenty of fine yellow silt was carried to the southwest by wind, and its deposition formed the loess plateau. In this manner the loess area was formed adjacent to the Mu-us Sandy Land. The topographical change from sandy land to loess hills is not gradual but drastic in geographical scale.

The Great Wall ranges on the border between the sand and loess areas (Figure 14.2). This shows that soil-type determines lifestyle and livelihood. Historically, to the north of the Great Wall Mongolian pastoralists rode horses

Figure 14.1 Location of the Mu-us Sandy Land

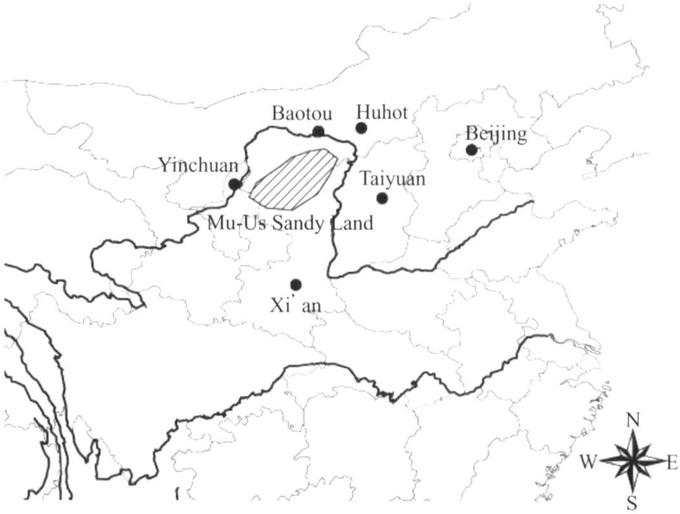

Figure 14.2 Landsat MSS image of the Mu-us Sandy Land (August, 1978)

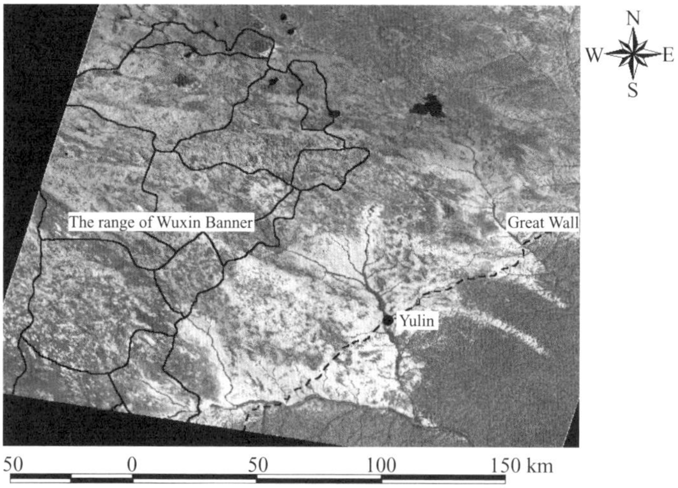

Note: White tone shows the area of active sand dune. Solid lines show the borders of the villages in Wuxin Banner. A dashed line shows the location of Great Wall. The range of Loess Plateau is visible in the south of Great Wall.

Photo 14.1. The Great Wall fragmented by gully erosion on loess hills

and raised sheep on the grassland. To the south of the Wall, the agricultural Han farmers cultivated crops on winding loess hill slopes. The northernmost area is in the Inner Mongolian Autonomous Region, and the southernmost area is in Shanxi Province.

Land productivity differs markedly north and south of the Wall. In the sandy land, crop growth is very poor due to the soil sterility and low water-retention capacity. Pastoralists utilize sparsely distributed plant biomass with animal husbandry. The population density has been kept low in this area. In the loess plateau, crops grow well due to the fertile soil which covers all the hills. The population density In the south is thus much higher than that in the north.

Contrastive ecological situations made quite dissimilar cultures. Mongolian people traditionally lived nomadically using Mongolian tents (*ger*). The region was governed by Mongolian royalty with a loose tributary to the Chinese empire. The pastureland was owned by several royal families, members of the nobility and temples. Herds and flocks were provided by these owners to the people, who were allowed to pasture animals. Livestock is the most important property for Mongolian people. The nomadic life has been lost and people are settled. Their houses are, however, dispersed and they do not aggregate to form villages. Their main diet is dairy products, and grain is purchased from Han people. They speak the Mongolian language and practice Tibetan Buddhism.

Conversely, Han people have settled in villages and cultivated millet and beans. They leased small areas of farmland from landlords before liberation where they cultivated millet and beans. Land is the most important property for Han people. They speak the Chinese language and worship the dragon and other spirits. The dragon is the symbol of rain which is the most desired natural phenomenon for dry-farming people.

In the boundary towns, markets were held for the exchange of Mongolian goods like horses and wool, and Han products such as grain and tea. Yulin city was the most famous market town in this area (Figure 14.2). In the boundary area, grassland is used by both Mongolian and Han people. The croplands of the Han territory were allowed to be grazed by Mongolian livestock during winter.

Nomadic farming—Invasion of loess farming

The landscape of Mu-us Sandy Land can be classified into four land-types (Figure 14.3). The first classification is a hard sandstone hill which is formed of the bedrock of the sandy land. Since the stone was formed in the warmer climate in the Cretaceous period, the iron content of the stones was oxidized to present the red color stratum. The soil cover is thin and sandstone is often exposed. Vegetation is characterized by xerophytic grasses and shrubs like *Stipa*, *Lespedeza*, *Caragana* and white *Artemisia*. This land-type is typcially used as extensive grazing fields for animals, because the land productivity is low.

The other land types are Quaternary sediments. The second form is a meadow steppe, where moisture is provided from groundwater. Vegetation is composed of mesophyte and hygrophyte. Plant productivity is high and the meadow is consequently used as the major pastures of animal husbandry. Shrubby *Salix* is sometimes developed and is called the 'willow stream scrub'.

The third and fourth land-types are sand dunes. Sand dunes are classified into active and fixed dunes according to sand mobility. An active sand dune in this region moves one meter southeast every year, propelled by strong spring wind. Active sand dunes are exposed without vegetation cover or are

Figure 14.3 Landscape structure and geological base of the Mu-us Sandy Land

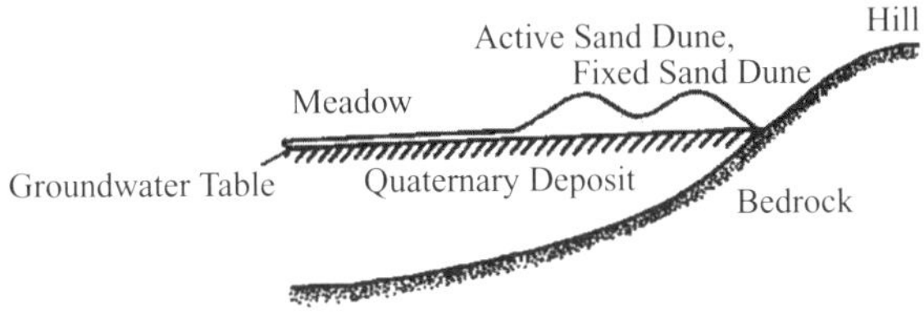

Source: reproduced from Kobayashi (1990).

Photo 14.2. Hard hill composed of Cretaceous sandstone

Photo 14.3. Meadow as pasture for animal husbandry

Photo 14.4. Active sand dunes

Photo 14.5. Fixed sand dune covered with Sabina scrub

covered sparsely by psammophytic annual plants or rhizomatous species. Fixed sand dunes are covered by loess soil and dominated by *Artemisia ordosica* or other shrubs. Though plant productivity is not as high as in other land types, these areas are used as winter grazing fields.

The relationship between active and fixed sand dunes is dynamic (Figure 14.4). Even though the landscape looks like desert, ecological succession operates here. The major factor preventing the establishment of plants is

Figure 14.4 Ecological succession of vegetation and land on sand dunes

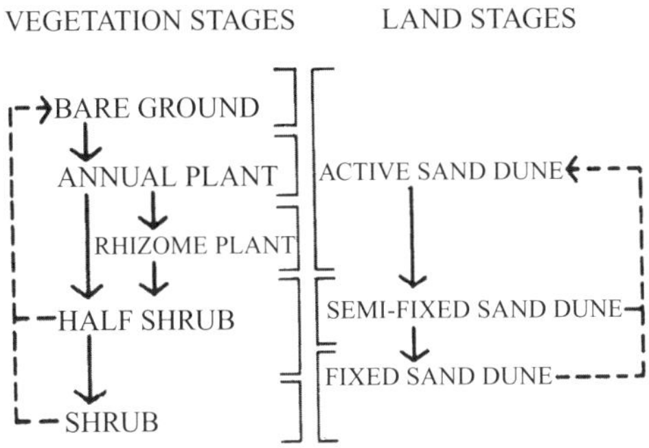

sand erosion. If sand mobility was reduced, plants would colonize the dunes. *Artemisia ordosica* is observed only when the annual change of sand surface is less than twenty cm. Under less mobile conditions, the development of vegetation is accelerated and the topsoil layer is formed. Conversely, if vegetation cover is obliterated on the dunes and the topsoil layer is disturbed, the sand underneath is again mobilized by strong winds.

Sand mobility is closely related to wind velocity and influenced by ground surface condition. If there is no groundcover, sand drifts in an exponential relationship with wind velocity. Most sand, particularly large particles, drift in the surface layer on the ground. Consequently, even tiny vegetation has the effect of reducing sand mobility. The technology of sand dune stabilization, termed 'grass quadrant', uses this physical principle. On the contrary, fine silt flies through the air. Therefore, active sand dunes consist of pure sand of large particles. If sand mobility is decreased, silt deposits on the ground surface and a thin loess layer is formed.

Through the middle years of the Qing Dynasty, nomadic pastoral life had broadly developed over this region. Since the early twentieth century, the Qing Dynasty changed the former pastoral policy and introduced Han people north of the Wall. The colonists brought intensive loess farming methods. Loess farming is characterized by tillage for the infiltration of rain and preservation of soil moisture. At first, these farmers removed vegetation on fixed sand dunes and ploughed the thin loess layer. They sowed millet seeds in spring and left the field to grow. They returned in autumn and harvested the sparse yield nourished by the sterile soil. Next spring, strong winds blew on the denuded ground and easily eroded the plowed topsoil layer. The hidden sand in the under-layer was consequently exposed and drifted again. The farmers abandoned the eroded fields and moved on. Desertification was thus rapidly developed by the shifting patterns of such nomadic crop-farmers.

Photo 14.6. Topsoil layer of fixed sand dune. 'Nomadic farming' was expanded, depending on this thin loess layer. Subsoil layer is composed of sand.

The early half of the twentieth century was a hard period for the Mongolian people in the complex political relationship between the Mongolian Royalty, Qing Dynasty, National Party, Communist Party and the Japanese Army. From 1901 to 1905, the counter-movement against the Han peoples' colonization was developed all over the Ordos region. The conflict was considered to be severe in Wuxin Banner, because it is the frontier to the Han world. Wuxin Banner is the local administrative region which covers most of the Mu-us Sandy Land.

In 1943, Mongolian soldiers rose against the governance of the army of the Republic of China under the Nationalist Party. The movement was suppressed and many Mongolians were killed in the campaign. In 1950, Mongolian soldiers again rose against the Communist Party's People's Republic. The last battle was carried out in the Mu-us Sandy Land and the Mongolian troops were annihilated.

Just after the establishment of modern China, crop fields were allocated to local people and Han farmer colonization decreased. During the period of the Great Leap Forward, however, many Han farmers again moved into the Mu-us Sandy Land. For example, the largest meadow steppe, Huangtologai, was a common grazing pasture for Mongolian and Han people until the 1940s. In 1949, a few families colonized the area from Yulin City and initially began farming. From 1957 to 1958, the Baihe reservoir was constructed on the wide valley. During the Great Leap Forward period, many Han families colonized the reservoir and formed a 'people's commune'. Now, most of the plain is owned by Han people. Huangtologai is a strange name, because 'huang' means yellow in the Chinese language and 'tologai' signifies a round hill in the Mongolian language. Just above the plain, there is a hard hill called 'Ulantologai' which means a red hill in Mongolian. The name Huangtologai might symbolize the history of 'Hanization' of this wide meadow under a hard hill.

Overgrazing: the modern process of desertification

Just after the establishment of the People's Republic of China, the human population was less than 30,000 and the total population of domestic animals was less than 25,000 in Wuxin Banner (Figure 14.5, 14.6). Since the area of Wuxin Banner is 30,000 km^2, human population density was not higher than one person per km^2.

Both the human and domestic animal populations significantly increased after that time. Though the movement of people was strictly regulated after 1960 and birth rates have been suppressed since 1972, the human population of the 1990s was three times larger than that in the 1940s. The total number of animals in the 1990s was also threefold that of the 1940s, based on government statistics. As discussed below, the actual number of animals might be larger than these data represent.

Land ownership here has changed radically in the last sixty years. Before the 1940s, the land was owned by landlords. Pastoral people borrowed herds, and some led a nomadic life (Editorial Committee of Regional History of Yikezhao League 1994). In 1952, the Cooperative System was started, succeeded by the People's Commune system in 1958. All land was owned by the commune and animals were provided by the commune during this period. Nomadic living was prohibited in the late 1950s by the government, and all pastoral people were settled. But the pasture was not partitioned during this period, and people were able to herd animals over a wide area.

In 1981, the People's Commune system was replaced by the Production Contract and Responsibility System, in which animals and land were allotted to each family. Land ownership actually became private. Though there were some troubles regarding land allotment, a Certificate of Pasture Range Contract was made for each family in 1997, securing the right to use an area of land for forty years. This contract system also incorporates a roll of land

Figure 14.5 Yearly population changes of Wuxin Banner

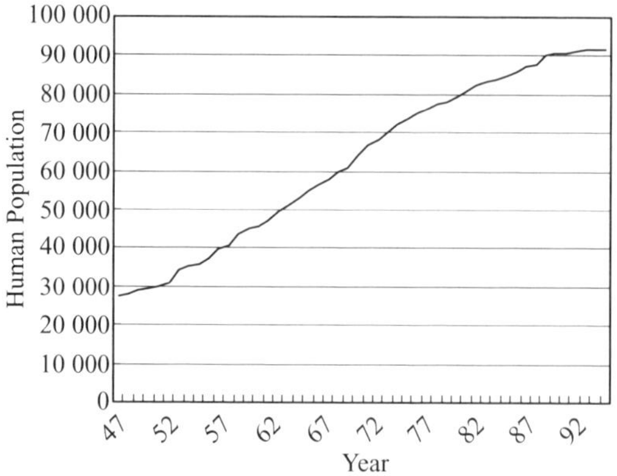

Source: produced with local government statistics data from the Statistics Bureau of Yikezhao League, Inner Mongolia, 1997.

Figure 14.6 Yearly animal population changes of Wuxin Banner

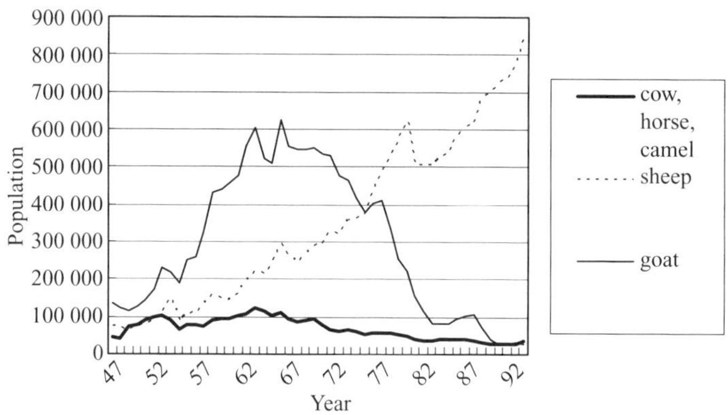

Source: Based on local government statistics, produced with the data of Statistics Bureau of Yikezhao League, Inner Mongolia, 1997.

register for taxation purposes. Each family was expected to manage their land in a sustainable and effective manner, and the fencing of pasture was encouraged by the government. With the erection of fences, animals no longer grazed widely. The current style of animal husbandry should be recognized as quite different to that practiced in traditional Mongolia.

Desertification increased after the establishment of the People's Republic in spite of the efforts of people and government to combat it. In the past, the major factors underpinning desertification were attributed to the introduction of inappropriate agricultural practices (see the Department of Geography, Peking University et al. 1983). Some alternative opinions have been proposed as to the causes of current desertification such as climate change, fuel gathering and over-grazing. In particular, over-grazing was suspected as a major factor, because the animal population was significantly increased as shown above. However, critical values regarding the carrying capacity of grassland differed markedly among scientific reports. For example, Li and Gao (1990) reported that the carrying capacity of grassland in the Mu-us Sandy Land was 0.3 to 1.0 head/ha, on the basis of the relationship between grassland productivity and the diet of a sheep. Ohkuro and Nemoto (1996) reported four head/ha, based on experimental data in the Kerqin Sandy Land in eastern Inner Mongolia. In order to study the factors behind modern desertification, a comprehensive study was planned with satellite remote sensing data and field surveys.

Through the preliminary survey of the first three years from 1997, we gained some knowledge. The analysis of satellite imagery and the results of extensive field surveys suggest the following findings:

1. in order to extract the trend of desertification, it is better to pay attention to fixed sand dunes than to meadows, which are easily affected by the fluctuations in precipitation
2. there is a general decreasing trend of observable vegetation on the side of the Inner-Mongolia Autonomous Region in contrast with an increase on the side of Shanxi District
3. though many sites with the characteristics of desertification are scattered in Wuxin Banner, the distribution of desertification areas cannot be explained by any simple factors such as topography or vegetation type.

These points suggested that there are complex anthropogenic factors underpinning desertification in this case. A field survey was conducted in 2000 involving actual visits to the site where drifting sand was observed in satellite image analysis, using a Global Positioning System (GPS). During the fieldwork, every landowner was interviewed regarding information about demography, economy and other household matters. Local statistics about land use and human and animal populations was collected.

LANDSAT/MSS data of August 17, 1978 and LANDSAT/TM data of September 5, 1996 were taken and correlated with GPS data. Land-cover type was classified into six categories by unsupervised methods. Two land-cover images were compared and land cover change was identified. Areas where the vegetation-covered land had changed to bare land were judged as desertified locales. Areas where the reverse scenario had occurred were deemed vegetated areas.

A 1:300,000 satellite map was edited from the LANDSAT/TM image with latitude and longitude meshes. Fourteen field survey sites were selected based on the various levels of desertification, and were plotted on the satellite map. In the summer of 2000, the study sites were visited accurately by car and on foot using the satellite map and GPS. Vegetation and soil were surveyed in each site. Twenty-one families were interviewed about human and animal population, land types and areas, income, fuel type and well depth. Complex desertification factors were analyzed by the comparison of these data.

Land area was quite different among interviewed families (Figure 14.7). The land area for crop farming was smaller than two ha for Han families 18a, 19a, 20 and 21. The cropland area per person was about 0.3 ha, including children. This land was surrounded by windbreaks and was seldom eroded.

The land area of pastoral people was large and widely variable, from thirty to 600 ha. The grassland area per person also varied from four to ninety ha. Land was reallocated after the abolition of the people's commune in 1978, but this distribution seemed to be uneven. Land change was also at an advanced stage after the reallocation. Distributed land area was not apparently correlated with desertification status (Figure 14.7). However, large areas of grassland often appeared at the sites of light desertification.

Results of a field survey showed that animal density in the heavily desertified area was higher than that in stable land. In particular, the presence of goats was eminent in the severely denuded sites (Figure 14.8). Other animals usually grazed on grasses, whereas goats browsed on even the shoots and roots of shrubs. Therefore, goat-attacked vegetation was changed to land cover dominated by annual plants. Regarding the results of an extensive survey, animal grazing was found to be the main factor in desertification and the role of goat is particularly important for soil erosion.

The landscapes around family 4 of Figure 14.8 illustrated the development of desert-ification. Though *Artemisia ordosica* was dominant here, its community was degenerated due to grazing within the fences of this family's land. Dead clumps were obvious and an annual herb, *Bassia dasyphylla*, which livestock avoid, was prolific in such places (Photo 14.7). Once shrubs have died, the topsoil layer of stable sand dune is apt to suffer erosion because of lack of soil-binding ability and the reduction of resistance to wind. Gully and sheet erosion were observed everywhere, and sand drift was also rife (photos 14.8 and 14.9).

Figure 14.7 Changes to sampled families' owned land area due to desertification status

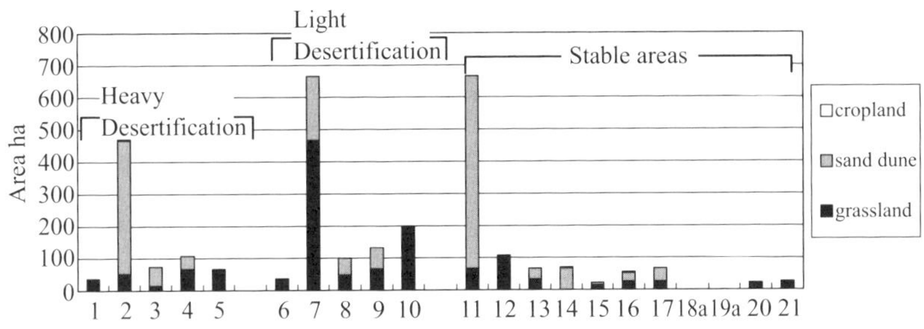

Source: Kobayashi et al., (2005)

Figure 14.8 Changes to sampled families' owned animals due to desertification status

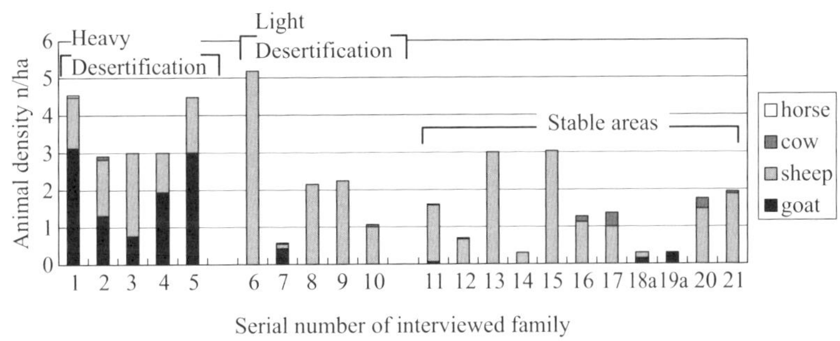

Source: Kobayashi et al., (2005)

Based on government statistics, the population of goat was supposed to be regulated, as shown in Figure 14.4. In fact, many goats were observed in our survey, and caused serious degeneration of vegetation and soil erosion.

Figure 14.9 shows the result of interviews about the income of sampled families. In heavy and light desertification sites, the income of ten out of eleven sampled families was more than 10,000 yuan. In stable areas, the income of seven out of eleven sampled families was less than 10,000 yuan. Though income is primarily related to the area of grassland and cropland, economic productivity was higher in the desertification group. In such areas, high income was brought about by the high density of livestock. Since goat provides cashmere wool, the highest income can come from goat farming.

In fact, the local government had prohibited goat farming, but to little effect. Electricity had not reached most houses in this area even by the middle

Photo 14.7. Vegetation changes due to damage through animal grazing. The fence traverses across the photo. Vegetation differs between the front and back of the fence.

Photo 14.8. Gully erosion by water-flow occurring at the shrubby *Artemisia*-declined location.

Photo 14.9. Wind erosion developed at locations without vegetation and topsoil.

Figure 14.9 Changes to sampled families' annual income due to desertification status

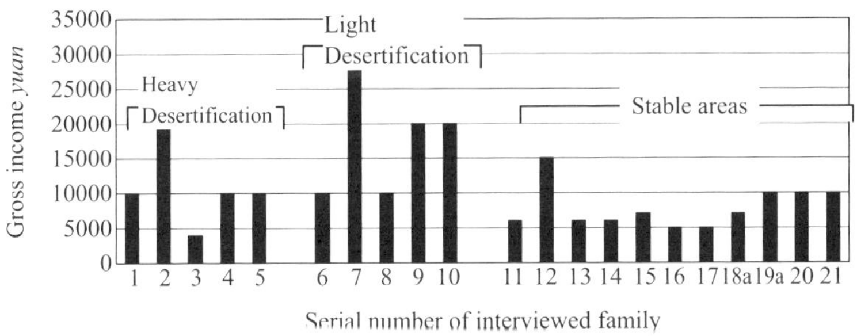

Source: Kobayashi et al., (2005)

1980s. In the 1990s, the Chinese economy was advanced significantly, and electricity lines were laid into local houses and major motor-roads were constructed. Now, radio and TV are available in many houses and people ride motor-bikes rather than horses. The people need currency to purchase such industrial goods and services, and consequently the number of sheep and goats increased in their own pastures. The annual income from a sheep was thirty *yuan* and that from a goat was fifty *yuan* in 2000. I saw an awarded 'advanced farmer' diligently raised the number of his animals, particularly goats. As a result, the land owned by the 'advanced farmer' was severely eroded.

The concept of appropriate animal population might not be traditional in Mongolia. Scientific records about traditional animal husbandry were not available for the Mu-us Sandy Land. The ecological report (Umesao 1953) about pasture usage in Chahal and Shilingol was used in this study. He reported that Mongolian people did not pay any attention to the management of grassland. They did not consider the carrying capacity of pasture. He concluded that the Inner Mongolian grassland was not saturated by animal population, because primary productivity was much higher than the consumption by animals in the 1940s. He insisted that over-grazing does not cause desertification in Inner Mongolia.

Since the current animal density might reach saturation point in this region, it is important to establish the modern idea of appropriate population, which might differ from traditional culture.

There were some environmentally conscious livestock farmers whose lands were not degraded in our survey. They told us that 'We do not raise goats, because goats deteriorate land'. Desertification was also not happening in Tuka *sum* (village in the Mongolian language). People said that the chief of the village strictly prohibits goat-breeding. People tried to change *Artemisia* scrub into grassland in order to raise cows in this village. These facts suggest that environmental education and strong leadership are effective to control desertification.

Desertification was not observable around the agricultural village of Nalinhe. Agricultural land use had a positive effect on the conservation and rehabilitation of vegetation cover. Recently, irrigation systems cover the agricultural villages where water is available from groundwater or streams, and temporal rain-fed fields have been changed to irrigation farms. Extensive farming styles have been changed to intensive ones in this area. Though the land is distributed to each farmer, the residual collective regulation of land use is functioning well. Around Nalinhe, there is a buffer zone between village and sandy land. The exploitation of new cropland is forbidden there and pasturing requires the permission of the community. The local government develops comprehensive policies with taxes, subsidies and volunteer activities to facilitate reforestation (Photo 14.10). Nalinhe operates as a stronghold of the counter-movement against desertification.

Changes and adaptations of the pastoral lifestyle under desertification

Into the new millennium, the Chinese government shifted the balance of national policy from economy to environment, along with the apparent

Photo 14.10. An old farmer cutting the leafy shoots of *Salix matsudana* in the windbreak of an agricultural village. The shoots were stocked and supplied for livestock as the forage in winter and spring.

economic developments. China Grassland Act was revised in the National Congress of People's Representatives in 2002. Due to this revision, the phrases 'improve ecological environment and conserve biodiversity' and 'sustainable development of economy and society' were inserted into Chapter One. Mindfulness of ecology was set as the most important issue in grassland use. In Chapter Forty-five, the following sentence was written: 'The nation determines the number of livestock for grassland due to the harvest of grass, and executes the system to maintain the balance between grass and livestock. The grassland service of the higher level of local government than county periodically evaluates the appropriate livestock-load on grassland according to its actual condition, using the standard livestock-load on grassland constituted by the Grassland Service of the National Administration Office. The People's Government of each class should take measures to prevent excessive livestock-load on grassland'.

Consequently, it was legislated that local people must manage grassland based on the livestock-load indicated by government. According to this rule, the prohibition of pasturage or the control of ranging was ordered for grassland where the quality of grass was degraded. Livestock tax was meanwhile abolished. These regulations were gradually introduced into local areas. If the period of pasture-grazing was restricted, forage was supplied by the government during stockyard feeding. Migration of pastoral people to new farmland was recommended in severely desertified areas in order to restore grassland.

In 2006, I revisited Wuxin Banner in order to understand the changes to people's lifestyles and the desertification status according to the change of environmental policy in China.

As a result of a survey, the regulation mentioned above was partly introduced in Wuxin Banner, but most areas remained as before. In the following paragraphs, the adaptation of local people to desertification and social changes are discussed, using four examples of pastoral families.

The most remarkable sign of desertification was seen in Ulantologai during fieldwork in 2000. We walked ten km to the center of the desertified area from the road with GPS. Ruins of houses were often observed along the path in the flat plains overtaken by active sand dunes. We visited the house of the pastoral family of AT (family 2 in figures 14.7, 14.8 and 14.9). AT was thirty-six years old and had six family members in 2000.

He mentioned that this place was named as Baiyingol, which means 'beautiful stream' in Mongolian. According to AT, there used to be many beautiful meadows with willow thickets, and even wild animals such as fox and wolf had lived there thirty years ago. He insisted that the vegetation was quite degenerated in the 1980s. AT had seventy goats, eighty sheep, five cows and one horse on fifty ha of grassland. Four-hundred and fifty ha of his land was sand dune. The population density of his animals was three head/ha, and the proportion of goats was high.

In 2006, the desertification status of Baiyingol was further advanced. A herd of goats was observed on the way to the house of AT. AT's livestock was left in to pasture, and we could not see and interview him again. One sterile goat was tied to a pole in the stockyards and a comb for cashmere was hung on the fence. This shows that goats were still kept by him. Although the same house was there, some of the window panes were broken. This area seemed impossible as a winter dwelling, and could only be used as a summer-camp for ranging.

We revisited BT (of family 7), whose income was highest among surveyed families in 2000 due to the harvest of natural *Ephedra*. His land area was

Photo 14.11. Goat pasturage was still extensive in the sandy land of Baiyingol in 2006.

Photo 14.12. The landscape of the hill-slope of BT's land in 2000.

also largest at 670 ha, which covered one hill where natural *Ephedra sinica* and wild leek grew. Although he kept 200 goats and fifty sheep, the livestock density was as low as 0.4 head/ha due to the broad pasture. His land was classified as being under light desertification.

E. sinica is a leafless xeric shrub growing on a sandy habitat. It is a medicinal plant and provides the material of ephedrine, which is a sympathomimetic agent and is used as a medicine for asthma and a cold remedy. Medicinal agents are contained in its stem. People harvest these with a scythe in winter, when soil is frozen and stems can be easily cut off. Since *Ephedra* propagates from rhizomes, new stems are regenerated in spring.

Collection of wild *Ephedra* seemed to be sustainable in 2000. However, vegetation decreased and desertification was expanded in 2006. BT increased the harvest of *Ephedra* from twelve to twenty ton/year, because the brokerage price of *Ephedra* declined from 1.75 to 0.7 *yuan*/kg. The spread of *Ephedra* cultivation in farmlands reduced the price of the wild plant. BT stopped goat breeding to minimize the damage of *Ephedra*. The increased consumption of *Ephedra* depleted vegetation cover.

BT maintained his income. He installed an electric generator and dug a 150 m deep well. He began cultivation of corn, sorghum and potato and irrigated these with an electric pump. Of course, TV and a gas-oven were furnished in his house. Even his traditional churn-staff was replaced by an electric stirrer.

HB's land (of family 4) was located on the hilly area, where sand drift was prominent in dry conditions. In 2000, he raised 160 goats and ninety sheep on the sixty-seven ha of grassland. Livestock density was four heads/ha, therefore his land suffered from grazing pressure. HB mentioned that desertification had become obvious since the 1980s, when the Production Responsibility System was adopted.

By 2006, large sand dunes emerged around HB's house and desertification further developed. During the six years prior, he expanded his pasture on another 107 ha of flat plain, and increased sheep numbers to 260 and goats to 190. Livestock density decreased down to 2.6 head/ha. HB showed us two Certificates of Pasture Range Contract. He bought a new one for 55,000 *yuan* with the right of land use for thirty years. Consequently, HB's income increased twice to 20,000 *yuan*. His strategy to survive in the less-productive sandy land condition was the extension of his business. But it was not enough to restore land condition.

CB's land had been owned by his younger brother in 2000 (family 5). His brother kept 200 goats and 100 sheep, but his land was completely eroded and rocks appeared on the slope of hills caused by serious damage of animal

Photo 14.13. Photo taken from the same position as Photo 14.12 in 2006. The degradation of the vegetation is apparent.

Photo 14.14. *Ephedra sinica*—a wild medicinal plant in the sandy land. Its price declined from 2000 to 2006.

grazing. Since there were few grasses except poisonous plants, he leased the neighboring grassland for pasturage. After we left the location, CB's brother gave up livestock farming and moved to a town.

CB succeeded the bare rocky land from his brother, moved there, and reduced the number of goats to sixty. He stopped ranging on his land, and employed a man to herd his goats to pastures in distant mountains, where his brothers share farm. The number of sheep was increased to 200. However, CB's main income did not come from sheep and goats, but from pig breeding. He had five sows, and transported seventy-five newborn piglets directly to big cities such as Baotou or Wuhai outside Wuxin Banner every year. CB earned 70,000 *yuan* from this business. Though his feed (mash) costs were high, his total net gain reached 30,000 *yuan* per year.

Vegetation was partly restored on the eroded rocky land because of less grazing. Even *Caragana* saplings, which is a natural dominant vegetation around this area, were established. CB overcame desertification through piglet production as an alternative profitable business.

Around the same area, another livestock business was introduced. Dairy farms with a corn field by well-irrigation have increased near the main road. Local government promotes dairy farming as a sustainable form of animal husbandry. A few small dairy factories have also been built on locations convenient for transportation.

By 2006, the control of ranging had not yet been introduced to Wuxin Banner. Since the grassland of this region is usually classified as temperate desert steppe, which has 0.36 head/ha carrying capacity of livestock according to figures shown in China Grassland Resources (1996). Twice to seven times this number of livestock is actually ranged, and the carrying capacity might not bee agreed upon by local people. Particularly in the desertified grasslands, pasturage should be prohibited. Though migration should be recommended in such cases, the construction of new settlements is difficult, because appropriate available space is sparse in the sandy lands. Dairy farming with deep wells is expected as an alternative pastoral lifestyle. But the distance to consumption areas from dairy factories is at least 100 km, and most houses of livestock farmers are inconveniently located away from motor roads. The future of this business model might, therefore, be questionable. Such situation makes the execution of ecological policy difficult.

Sneath (2000) reported that pastoral life in Inner Mongolia was still maintained under the governance of the People's Republic even after the revolution. Recently, though, the Mongolian lifestyle has been drastically changed through adaptation to economic development and desertification.

Photo 14.15. Completely denuded land by goat-grazing in CB's land in 2000.

Photo 14.16. Photo taken from the same position as Photo 14.15 in 2006. The land was revegetated by the discontinuation of goat-ranging.

Some have bought land to manage a pastoral business. Others have chosen crop and dairy farming. Still others have begun to breed the Han Chinese animal, the pig, as a sustainable business. Most people have now abandoned the use of horses. Though the National Government started the social and ecological policy, some local people independently adapt their economic and ecological surroundings. Further attention should be paid to the future of pastoral culture, society and ecology.

References

Department of Geography, Peking University, Natural Resources Study Committee, Chinese Academy of Sciences, Lanzhou Desert Research Institute, CAS, Lanzhou Glacier and Frozen Soil Research Institute, CAS (1983), *Natural Condition and the Improvement of the Mu-Us Sandy District* (in Chinese), Peking: Science Press.

Department of Livestock Veterinary, Ministry of Agriculture, People's Republic of China and Central Station of Livestock Veterinary (1996), *Chinese Grassland Resources* (in Chinese), Peking: China Press of Science and Technology.

Editorial Committee of Regional History of Yikezhao League (1994), *History of Yikezhao League Vol. 2* (in Chinese), Peking: Modern Press.

Kobayashi, T. (1990), 'Origin and land classification of the Mu Us Shadi Desert in China' (in Japanese), *Japan Journal of Revegetation Technology*, 15: 43–57.

Kobayashi, T., S. Nakayama, L.-M. Wang, G.-Q. Li and J. Yang (2005), ëSocio-ecological analysis of desertification in the Mu-Us Sandy Land with satellite remote sensingí, *Landscape and Ecological Engineering*, 1: 17–24.

Li, B. and Y. Gao (1990), 'Grassland resources in Ordos Plateau', in Yikezhao League Team, Remote Sensing Survey Project of the Grassland Resources in Inner Mongolia (ed.), *Studies on Natural Resources and Environment of Ordos Plateau, Inner Mongolia* (in Chinese), Peking: Science Press: 135–152.

Ohkuro, T. and M. Nemoto (1996), 'Vegetation degradation process caused by sheep grazing at sandy grassland in semi-arid regions of northeast China' (in Japanese with an English summary), *Papers on Environmental Information Science*, 10: 31–36.

Sneath, David (2000), *Changing Inner Mongolia—Pastoral Mongolian Society and the Chinese State*, New York: Oxford University Press.

Statistics Bureau of Yikezhao League, Inner Mongolia (1997), *Bright 50 Years of Yikezhao League* (in Chinese), Peking: China Statistics Publishers.

Umesao, T. (1953), Bod and Stoss: The fragment of pastoral ecology (in Japanese), *Chiiki*, 8: 2–8.

15
The Socioeconomic and Environmental Roles of Bamboo in Rural China

Manuel Ruiz-Pérez, Fu Maoyi, Brian Belcher, Yang Xiaosheng, Benoit Mertens, Liu Hua and Xie Jinzhong

Introduction

Man can live without meat, but cannot live without bamboo

These words attributed to Confucius express the close relationship between the material and spiritual values of bamboo in China. Bamboo is ubiquitous as the raw material of a multitude of products used in daily life and as a source of symbolic, artistic and aesthetic inspiration (Williams 1941). Moreover, Chinese culture attaches great spiritual significance to bamboo. One of its key roles is as a mediator between the earthly world and the world of the ancestors—the smoke from burning yellow bamboo paper represents the contact and transmission of wealth between generations.

With a millenarian tradition and a well established cultivation practice,[1] China has the largest bamboo resources in the world. There are thirty-nine genus and an estimated 500 species (with some variations depending on taxonomic criteria), of which around 100 have economic use (Zhu, Ma and Fu 1994). Chinese large woody bamboo can be divided into three basic groups according to taxonomical, biogeographical, management and utilization purposes. These are monopodial bamboo (with a longitudinally growing rhizome), distributed in the southern part of the temperate and northern part of the subtropical regions of China, sympodial bamboo (with a cluster growing rhizome), distributed in the southern and western regions, and amphipodial bamboo (combining longitudinal and cluster growing segments in their rhizome), occurring between both previously mentioned areas (Figure 15.1).

Bamboo has two main and well differentiated uses[2]. Bamboo culms (stems) are used as a source of solid material and fiber with properties comparable to wood (for which it is a common substitute, both in solid and pulp forms). Bamboo is also a source of food, mainly in the form of bamboo shoots. This

Photo 15.1. Farmer with his bamboo culms ready to take to the market in Muchuan County, Sichuan

dual use and relatively short rotation period gives bamboo plantations an intermediate role between a forest plantation and an agricultural crop. As we shall see, this fact has major implications in the evolution of bamboo and the role it plays in farmers' household economy. We present here a synthesis of key findings obtained over a ten year collaborative project between the Research Institute of Subtropical Forestry of the Chinese Academy of Forestry (RISF-CAF), the Center for International Forestry Research (CIFOR), the China Institute of Resource Information Technology of the Chinese Academy of Forestry (CIRIT-CAF) and the Autonomous University of Madrid. The project has evolved through different phases from its initial stage of collaboration between RISF, CIFOR and the International Network for Bamboo and Rattan (INBAR) in one county, to research carried out in eight counties in three provinces along an east-west gradient, incorporating farmers' surveys, official statistics, forest inventories and remote sensing analysis.

Photo 15.2. Farmer bringing bamboo culms from her farm to a bamboo paper factory in Muchuan County, Sichuan

Recent evolution of bamboo resources in China

The last published forest inventory of 1998 records 4.2 million ha of pure bamboo forests and plantations, to which another three million ha of mixed and mountain natural bamboo stands should be added (State Forest Administration 2000). Significant amounts of woody bamboo[3] occur in seventeen provinces, with ten provinces having over 100,000 ha each. Four contiguous provinces (Fujian, Jiangxi, Zhejiang and Hunan) have more than half a million ha each, accounting together for over sixty percent of total bamboo plantations in China.

The bamboo resource base has increased steadily over the past two decades, both in surface area (31.6%) and in density of stands (40.8%) (Figure 15.2). The combined expansion of plantations and density has resulted in

Figure 15.1 Distribution of main groups of bamboo in China

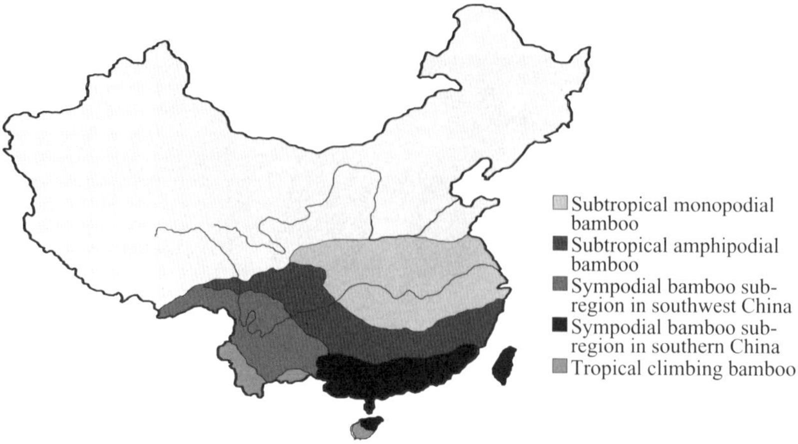

an exponential increase in output of bamboo culms and shoots (594% and 1048% respectively) that contrasts sharply with the stagnant and recently decreasing output of timber (Figure 15.3).

There are important differences in the rate of expansion of different types of bamboo. *Moso* bamboo (*Phyllostachys heterocycla* var. *pubescens*), a monopodial bamboo of high quality culms that represents over two-thirds of total bamboo plantations, has expanded at a slower rate (17.1%) than sympodial bamboo like *Bambusa spp., Dendrocalamus spp., Sinocalamus spp.* (with an 82.9% increase) in the two decades covered by the four forest inventories carried out by the Chinese government. This is the combined result of four factors:

1. lower cost to establish plantations of sympodial bamboo—they can frequently be planted from branches, whereas *moso* bamboo requires planting/transplanting root stock
2. earlier maturity of sympodial species—first harvest in three to four years after planting, compared to six to eight years for *moso* bamboo
3. new forest policies like The Natural Forest Protection Act and the conversion of agricultural land to forestry in steep slopes in the West, where sympodial bamboo are more common, that have created new opportunities and incentives for farmers to plant bamboo
4. increasing investment aimed at developing the Western regions of the country, including investment in bamboo-based industries that stimulate demand for bamboo raw material[4].

Figure 15.2 Evolution of bamboo area in China

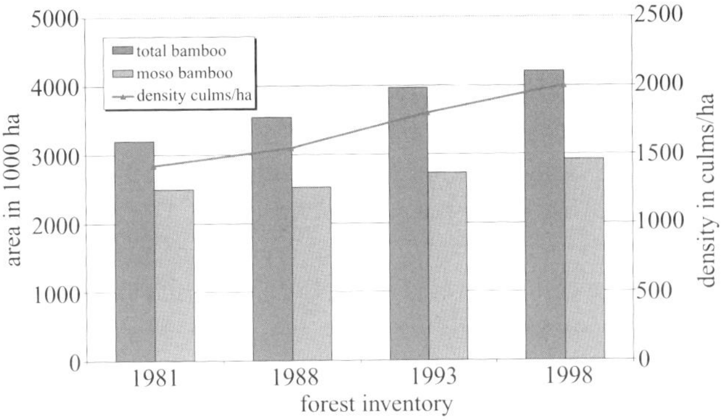

Figure 15.3 Timber and bamboo output 1980–2002

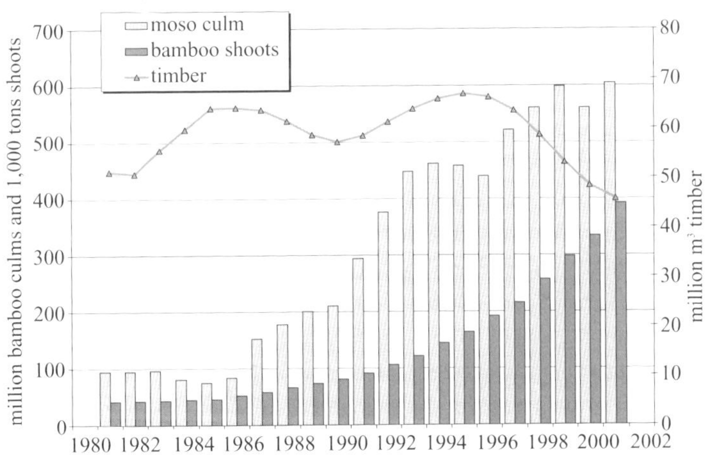

Consequently, the faster active expansion[5] of bamboo has taken place in provinces like Sichuan where sympodial bamboo can grow. A notable exception has been Yunnan that has seen a decrease of one-third in its bamboo area during this period. This decrease, which parallels the general decrease of forest resources in the province (Guo et al. 2002), has been recently redressed with the abovementioned new policies. Four Southwest provinces (Sichuan, Yunnan, Guangxi and Guizhou, the latter starting from a very low base) account for seventy-three percent of total new plantations established in recent years (State Forest Administration 1987–date).

Spatial patterns of bamboo expansion in three counties

Land-use change analyses based on LANDSAT images allows for a detailed understanding of spatial processes of bamboo expansion, and provides a good basis for comparison with the official statistics. The study, conducted in three counties (Anji, Pingjiang and Muchuan; see Figure 15.4) reveals a high increase of bamboo area in the late 1990s. During the first period of observation, the amount of land covered by bamboo did not change significantly at the county level (Figure 15.5). For the whole period, bamboo area increased from 514 to 690 km^2 in Anji (25%), fifty-seven to ninety-three km^2 in Pingjiang (38%) and from eighty-two to 140 km^2 in Muchuan (41%). Bamboo area, as measured by remote sensing, shows similar values than those from the statistics of the counties' forest bureau, except for Muchuan where the official statistics indicate a much larger bamboo area, a discrepancy that may in part be due to cloudiness in the images of this mountainous county.

Anji County is well connected to large external markets and big cities: the distance from Anji to Hangzhou (the capital of Zhejiang province) is about seventy km, and the distance to Shanghai (the largest Chinese city) is about 230 km. The transportation network within the county is well developed, with a road density of 0.34 km/km^2. About sixty percent of the county is covered by forests, of which about half is occupied by bamboo. Most forest areas are located at the periphery of the county, while a large central plain is occupied by agriculture. The bamboo expansion seems to be strongly shaped by the existing pattern of bamboo coverage. Indeed, bamboo expansion tends to be concentrated in limited regions, especially around already-established bamboo plantation areas: for the two periods of analysis, almost ninety-five percent of the new bamboo areas were created within one km of existing plots. Because of the high road density in the county, bamboo plots tended to be located near the roads. But in recent years, bamboo expansion has tended to take place in less accessible areas, at a longer distance from roads and at higher elevation.

Pingjiang County is located in Hunan province within the Centre region, 600 km west from the coast. The road network is less developed, with a length of about 600 km but a density of 0.16 km/km^2. More than eighty percent of the county is covered by forests, but only two percent is occupied by bamboo. The large majority of the bamboo plots in the 1980s as well as the new bamboo areas are located in the more elevated south-east periphery of the county. Bamboo areas are found generally in less accessible areas, far

Figure 15.4 Forest regions of China and location of the counties of Anji, Pingjiang and Muchuan

Figure 15.5 Changes in bamboo area in Anji, Pingjiang and Muchuan based on remote sensing analysis

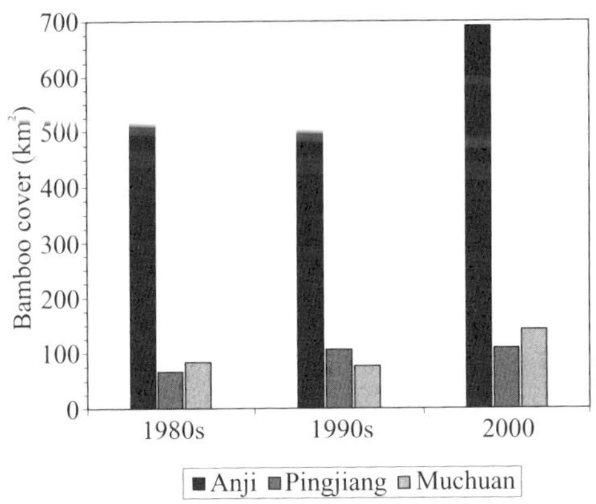

from the main roads and in places of high elevation. Bamboo expansion in this county is, however, slowly shifting towards more accessible areas.

Finally, Muchuan County is located in Sichuan province in the southwest region. This county showed the highest bamboo expansion rate among our study areas. The county is located near the border of Yunnan province which is actually the fastest expanding bamboo province. The road network is also less developed, with about 145 km in length, but a density of less than 0.1 km/km^2. About sixty-five percent of the county is covered by forests, of which five percent is occupied by bamboo. Mountains divide the county into two large watersheds in the southeast and the northwest. As in the other counties, bamboo areas tend to concentrate in these high elevation places. A large majority of the bamboo plots are located in the southern part of the county, which has better transport infrastructure and more bamboo processing factories. As observed in Pingjiang County, bamboo expansion tends to take place in more accessible areas, closer to roads and in places of lower elevation. A concentration of new bamboo plots is observed around two of the three main bamboo related factories in the extreme south of the county.

Spatial analysis reveals a distribution of bamboo near processing centres, with a contagious expansion around already existing plantations. However, while in Anji new bamboo plantations tend to be relegated to more remote areas, in Pingjiang and Muchuan the new plantations tend to occupy more accessible areas. This is due to different opportunity costs of land in a bamboo mature county like Anji as compared with counties like Pingjiang and Muchuan where bamboo covers a relatively small percentage of the territory and where its potential for expansion is still very large.

The context of bamboo expansion

Five main features characterize the bamboo sector in China, helping to explain its exceptional dynamism:

1. The dual use, as a supply of both timber and shoots in short rotation cycles with productivity (both for land and labor) that compares with agricultural crops makes bamboo an economically attractive option to farmers as a forest land-use.
2. The relatively small management unit required (with ten *mu*, equivalent to two-thirds of a ha, being a typical bamboo area managed by a family in our samples) means that bamboo can easily be adapted to the household responsibility system. In our sample of four counties bamboo

under the household responsibility system represents between eighty (Pingjiang) and ninety-nine percent (Muchuan) of total bamboo area.

3. Both bamboo culms and shoots are traded in open markets, being one of the first forest products in China to be exempted from State marketing board control. Market intermediaries, agreements between farmers and large bamboo processing firms and even incipient futures markets can be found in different regions.
4. The variable scale and diversity of processing units and the possibility of a degree of pre-processing by farmers makes local processing a viable option. Bamboo epitomizes the rural value-adding industrialization aimed at in many parts of China, while its diversity of uses allows for a certain degree of county level spatial specialization.
5. Expanding national, regional and international demand, both as substitute for timber and for some superior goods like bamboo shoots and bamboo flooring, has attracted national and foreign investment[6] seeking to develop new technologies.

In combination these features give bamboo an economic attractiveness and a flexibility that have made it a useful resource for adapting to and taking advantage of the opportunities offered by key policy reforms related to land tenure, commercialization, industrialization and exports.

The role of bamboo in the household economy

Bamboo has been characterized as an inferior good in the typical view of bamboo as the 'Poor man's timber' (Rao, Dhanarajan and Sastry 1987). However, contrary to this view, our research in China indicates that bamboo is frequently an average or superior commodity, able to adapt to new demands that increase with income (Ruiz-Pérez et al. 2003). Modern processing technologies have been developed to produce a remarkable variety of new products. From resistant plybamboo to attractive bamboo flooring, from medicinal bamboo extracts through to bamboo-based tourism, a thriving market has offered farmers attractive income opportunities. Research on changing sources of household income indicates that bamboo is generally a good option, with an increasing number of farmers deriving more income (in constant terms) from bamboo between 1985 and 2000 (Figure 15.6). The process of bamboo expansion is twofold:

1. expansion to new counties in provinces that formerly had little bamboo (as in the case of Guizhou);

Figure 15.6 Changes in the distribution of bamboo income in 1990 and 1995

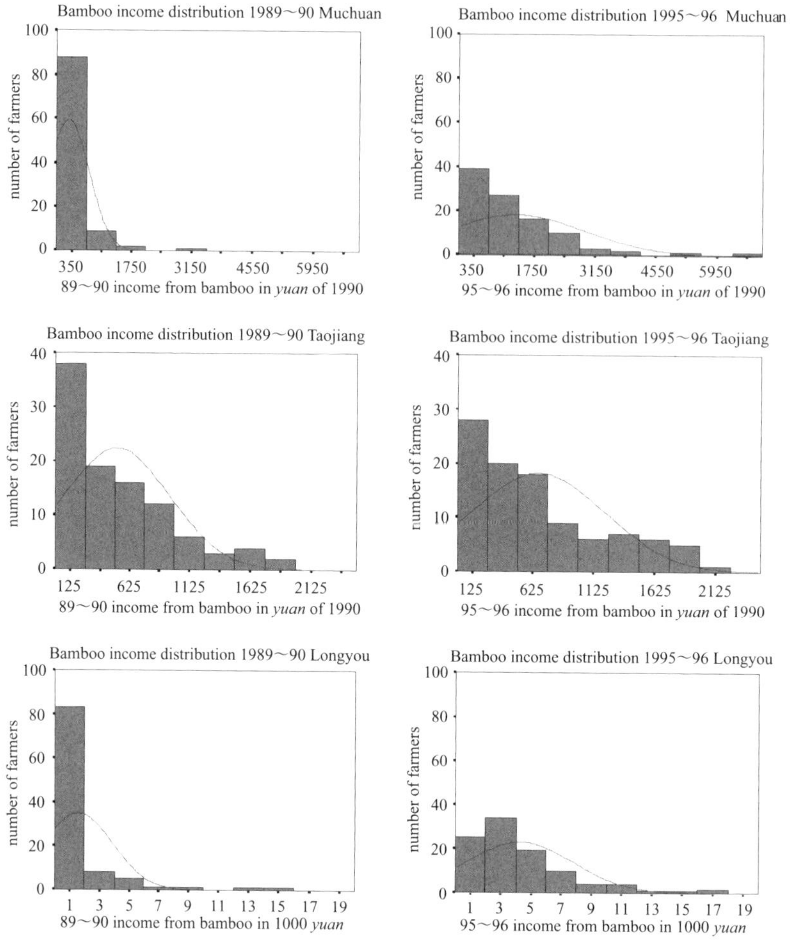

2. expansion in counties and provinces that had a tradition of bamboo, and where more farmers are cultivating bamboo.

Our research suggests that bamboo offers differential benefits to different types of farmers depending on local conditions and level of development. We propose that in early stages, when bamboo is developing fast, higher income farmers can derive proportionally[7] more benefits from bamboo than middle or low income farmers because they are better placed to take advantage of new opportunities. As the county and province develops and bamboo nears a plateau in its expansion, middle income farmers tend to get proportionally more income from bamboo.

Photo 15.3. Weighing bamboo culms in a traditional bamboo market in Daoming township, Sichuan

In a rapidly expanding economy such as China's, bamboo (or other natural resource-based activity) would become more important for low income farmers when the level of development is already so high as to offer more attractive opportunities (like off-farm income) or when bottlenecks (like poor infrastructure and general levels of development that preclude an associated industry) put a brake on any potential expansion of bamboo. Therefore, the 'Poor man's timber' and the associated view of bamboo as beneficial mostly for poor farmers is a particular, not very frequent case in China's bamboo regions.

We have expanded this proposition from the farmer to the county level. Analysis of the percentage of county-level processing, industry output and employment indicate that bamboo has played a significant role in the early stages of industrialization and development (Table 15.1). However, as the

Photo 15.4. Old farmer weaving bamboo baskets for fruit packing and transportation in Taojiang County, Hunan

Table 15.1: Increase in bamboo local processing, industry output and number of workers

	% increase in local processing (85–98)	increase bamboo industry output in constant yuan (85–97)	increase bamboo industry workers (85–97)
Anji	48	23.1	5.6
Longyou	61	3.7	5.0
Pingjiang	-18	0.7	2.2
Taojiang	61	124.7	15.7
Changning	65	1.9	2.4
Muchuan	0	9.2	3.7

Source: County level statistics, several years

Photo 15.5. Moso bamboo (*Phyllostachys heterocycla*, var. Pubescens) bordering rice fields in Jishangvillage, Longyou County, Zhejiang

county develops, better options appear, and bamboo industry opportunities and their relative wages decrease (Figure 15.7). Bamboo is a good foundation for rural development, remaining as a well established activity, but after a point the more attractive industries and incomes will have to come from other leading sources.

Some environmental aspects of bamboo plantations

Artistic representations, a stereotyped association with some emblematic conservation animals like Pandas, and a long history of cultivation have incorporated bamboo plantations in the idealized Chinese landscape. However, little is known about the environmental effects of bamboo

Photo 15.6. Moso bamboo (*Phyllostachys heterocycla*, var. Pubescens) expanding in a conifer plantation in Wu Chun village, Longyou County, Zhejiang

Figure 15.7 Relationship between per capita income and relative ratio of bamboo industry wage

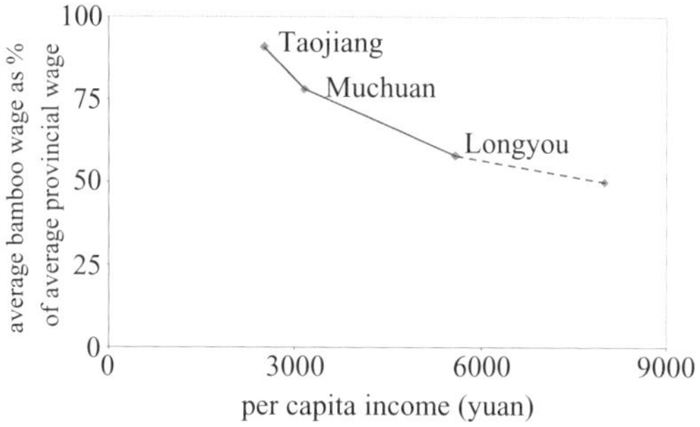

plantations. As a crop, bamboo has different management intensities, from rather extensive systems based on natural expansion of bamboo within mixed forests to very intensive monoculture plantations that require regular tillage and heavy fertilizer and pesticide inputs.

As with most other agricultural and forest land uses, bamboo plantations involve some trade-offs between productivity and environmental values. The proper assessment of these trade-offs has to be done in comparison with natural forests but also with alternative land uses with which bamboo has to compete. We have therefore compared two levels of bamboo management intensity with natural broadleaf forests[8], conifer (Chinese fir) plantations and economic forests (mostly tea and oil tea plantations). The study, which is being conducted in four counties[9] (Fuyang, Anji, Pingjiang and Muchuan), assesses the biodiversity and biomass of bamboo areas and its corresponding potential for carbon sequestration, in addition to water run-off and soil erosion. We present here some preliminary results for Anji County.

Natural forests have similar species richness as bamboo plantations, and more than double the total number of species found in Chinese fir plantations (see Figure 15.8). However, the composition of species varies, with fewer trees and more common herbaceous species as we move along the gradient from broadleaf to intensive bamboo plantations (Figure 15.9). Bamboo plantations contain between one half and one seventh the tree biodiversity found in natural forests.

Figure 15.8 Species richness in each form of land-use in Anji

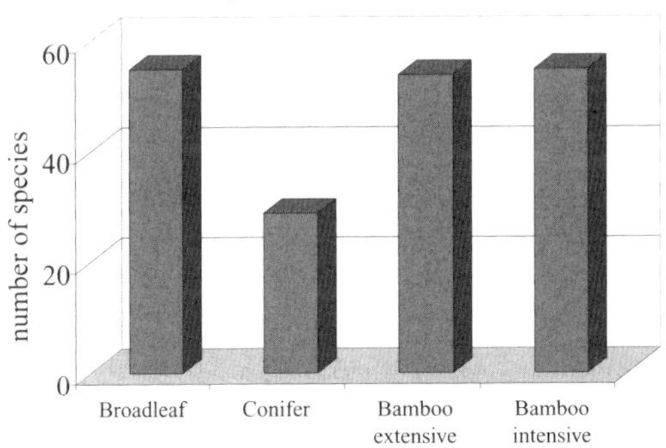

Figure 15.9 Relative abundance of plant groups by type of land-use in Anji

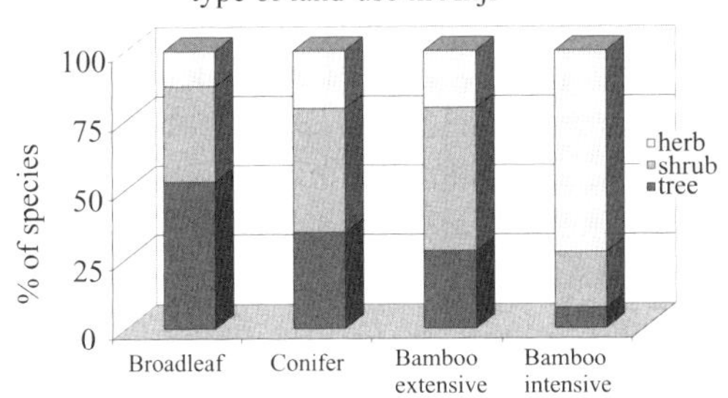

Total dry biomass is also highest in natural forests, reaching an average of 172.5 tons/ha (Figure 15.10). Anji has very intensive bamboo plantations, with a total biomass of 84.0 tons/ha, higher than the biomass of Chinese fir plantations with 72.1 tons/ha. Finally extensive bamboo, with 37.8 tons/ha, have significantly lower biomass than the other two main types of plantations. As we can see, the differences in tree biodiversity in Anji are paralleled with equivalent differences in biomass. None of the alternative land uses reach fifty percent of the biomass of natural (even if disturbed) forests. There are trade-offs between biomass and biodiversity when we compare intensive versus extensive bamboo plantations.

Run-off and soil erosion series over a two-year period show that natural forests have the lowest run-off and loss of soil (measured in kg of soil per ha; see Figure 15.11). Chinese fir and extensive bamboo plantations yield significantly higher amounts of water due to a more open canopy cover and fewer, smaller trees. They also lose more soil, particularly the extensive bamboo plots, with twice the soil loss of natural forests. The loss of soil is even larger in the more intensive bamboo plantations and economic forests although these land-uses yield smaller amounts of water (comparable to the water yield of natural forests) due to their dense canopy cover that intercept a significant amount of rain. Soil losses in bamboo plantations and economic forests are between three and four times the amount of loss of natural forests[10]. Again, intensification of bamboo plantations has trade-offs between increased productivity obtained by intensification and losses of soil due to the frequent tillage of intensive bamboo management.

The new environmental dimension of the bamboo sector

In the context of the increasing demand for forest products, prohibition of cutting natural forests and plantations in watershed protection areas, and renewed emphasis of Chinese authorities on expansion and improved plantation management, the bamboo sector will continue growing in China. The expansion of bamboo production will continue as a combination of expansion of plantations and intensification to increase productivity.

The rate of expansion of plantations will vary depending on local environmental and socioeconomic conditions. Our survey indicates that the substitution of arable land (mostly in steep slopes) and less productive forest plantations by bamboo is more likely to occur in the Southwest as well as in some mountainous, poor provinces of the East like Anhui. Muchuan epitomizes this trend in the set of counties of our sample.

Figure 15.10 Biomass structure of different forest types in Anji

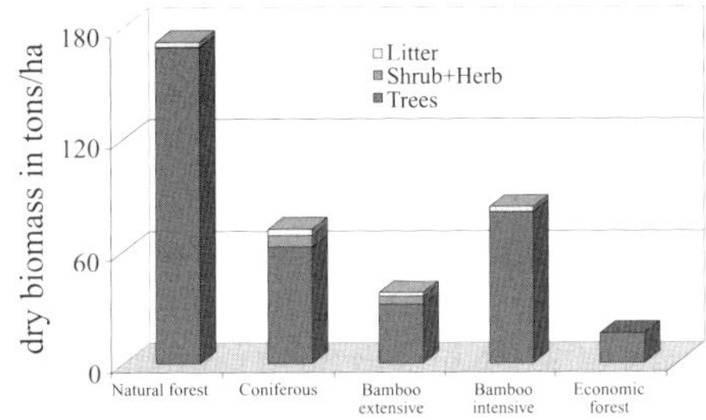

Figure 15.11 Run-off and soil erosion in Anji

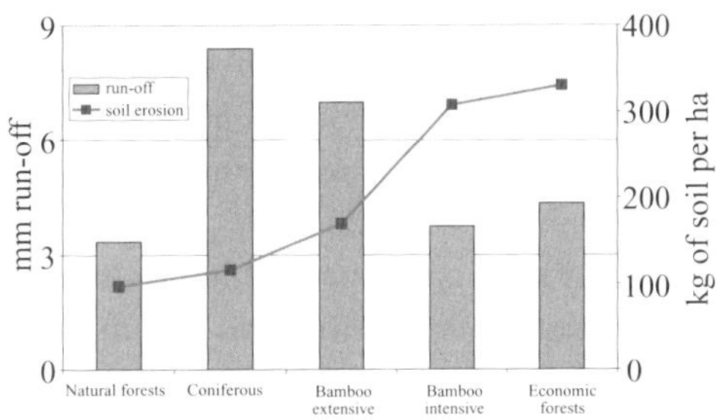

The new emphasis on environmental protection of forestlands by the Chinese Central Government reduces the possibility of replacement of natural forests or other types of plantations by bamboo, a practice that has been common for the past two decades. Counties like Anji with a mature bamboo sector where bamboo plantations have already taken most of the available land find it difficult to expand the current bamboo area. Because of that, intensification has been the path to increase output in this county. Anji has some of the most intensively managed bamboo plantations.

Local forest authorities are already aware of the trade-offs between intensification and environment described in this paper. Moreover, Anji has a well established reputation, being considered as one of the ten 'Bamboo Hometowns' in China, appealing to the bamboo aesthetic values deeply rooted in Chinese culture. The proximity to large and prosperous urban centers like Shanghai, Hangzhou and Nanjing, together with the vastly improved accessibility of the county, has attracted hundreds of thousands of tourists every year. Bamboo has entered into the tertiary sector economy in Anji with rural tourism programs like 'Visit the bamboo farmer home' and its popular Bamboo Botanical Garden. In order to maintain and improve the attractiveness and the environmental services offered by bamboo plantations, county forest authorities have initiated a process of controlling the negative effects of intensification, including incentives to farmers to reduce the intensity of existing plantations and to move from monocultures to mixed bamboo-natural forest plots (Ruiz-Pérez et al. 2001).

As part of the new emphasis on tourism and conservation, Anji County forest authorities have recently invested more than US$600,000 to build the China Bamboo Museum. In addition, an 18,000 ha National Forestry Park, mostly composed of bamboo, has been set up with the approval of the State Forestry Administration. Moreover, the China Great Bamboo Sea, with its 10,000 ha located in the eastern part of Anji County, is famous for having very big bamboo culms. All these projects are meant to protect the environment and to attract tourism.

Given the growing importance of the off-farm income in the farmer household economy and their diminishing dependence on agriculture and forest income, the incentives to move to extensive plantations (that at the same time help to save labor that could have good opportunities in other activities) seem to be playing a catalytic role in this county. Anji provides an exemplary but rare case of combining production and environmental protection that much deserves monitoring and emulation in today's China.

Acknowledgements

We acknowledge the European Union (through its core funding to CIFOR), the Spanish Cooperation Agency AECI and the Science and Technology Office of Zhejiang for funding of this research.

Notes

1. The first treatise on bamboo, *Chu P'u* [*Zhu Pu*] (Bamboo Chart), was written by Dai Kaizhi in the Jin dynasty (317–420 A.D) (Geng and Wang 1994).
2. There are a number of other uses, like medicinal, extracts to add flavonoids to beer, charcoal or litter for animal beds.
3. There are a large variety of thin-stemmed and herbaceous bamboo species that are economically less important.

4. This is the case of the fast expansion of bamboo plantations in the Southwest to supply raw material to a growing paper industry.
5. We can differentiate between an active expansion of bamboo, exclusively promoted by farmers, and a mixed expansion due in part to the aggressive growth of bamboo rhizomes that can colonise neighbouring lands. The second scenario is very common in monopodial bamboo, helping to explain the early use and current predominance of *moso* bamboo in Chinese plantations that also benefit from a less dramatic effect of synchronic flowering on *Phyllostachys* species (McClure 1993). However, as new demands appear, other species are being incorporated, thus reducing the rate of expansion of *moso* bamboo.
6. Foreign investment was more important in the early stages of the reform, mostly during the 1980s.
7. Measured as a percentage of total income coming from bamboo; in absolute terms as total income from bamboo richer farmers tend to get more bamboo-based income than do middle income farmers and these more than low income farmers.
8. They are disturbed or secondary natural forests. There wasn't any significant amount of primary, undisturbed forests left in the counties we studied for comparable altitudinal and soil conditions.
9. Part of the data are still being gathered, therefore the results showed in this section are provisional.
10. We do not include here the soil losses due to the establishment of new plantations. These can be very heavy when clear-felling and soil tillage is carried out in preparation for planting.

References

Geng, B. and Z. Wang (1994), 'Preface', in Zhu, Ma and Fu *A Compendium of Chinese Bamboo*, Beijing: China Forestry Publishing House.

Guo, H., C. Padoch, K. Coffey, A. Chen, and Y. Fu (2002), 'Economic development, land use and biodiversity change in the tropical mountains of Xishuangbanna, Yunnan, Southwest China', *Environmental Science and Policy*, 5: 471–479.

McClure, F. A. (1993), *The Bamboo, a Fresh Perspective*, Cambridge: Harvard University Press.

Rao, A.N., G. Dhanarajan and B. Sastry (eds) (1987), *Recent Research on Bamboo*, Proceedings of the International Workshop 'Recent

Research on Bamboo', Hangzhou, China, October 1985, Ottawa: IDRC.

Ruiz-Pérez, M., M. Fu, X.Yang, and B. Belcher (2001), 'Bamboo forestry in China: Toward environmentally friendly expansion', *Journal of Forestry*, 99 (7): 14–20.

Ruiz-Pérez, M., B. Belcher, M. Fu, and X. Yang (2003), 'Forestry, poverty and rural development: Perspectives from the bamboo sub-sector', in W. F. Hyde, B. Belcher and J. Xu (eds), *China's Forests. Global Lessons from Market Reforms*, Washington D.C.: Resources for the Future Press: 151–176.

State Forest Administration (various years, 1987–date) *Chinese Forestry Yearbook* (in Chinese), Beijing: China Forestry Publishing House.

State Forest Administration (2000), *Forest Resources Statistics of China (1994–1998)* (in Chinese), Beijing: Department of Forest Resources Management.

Williams, C. A. S. (1941), *Outlines of Chinese Symbolism and Art Motives*, Third revised edition, Shanghai: Kelly and Walsh Ltd.

Zhu, S., N. Ma and M. Fu (1994), *A Compendium of Chinese Bamboo*, Beijing: China Forestry Publishing House.

Index